Life Sciences Research and Development

Life Sciences Research and Development

Dynamics and Interrelations between Nature, Science, and Society
Kaushalendra Kumar Jha, PhD (Editor)
Michael Campbell, PhD (Editor)
2022. ISBN: 979-8-88697-053-1 (Softcover)
2022. ISBN: 979-8-88697-130-9 (eBook)

Cyanobacteria and Their Importance
Rajeshwar P. Sinha, PhD (Editor)
2022. ISBN: 978-1-68507-934-5 (Hardcover)
2022. ISBN: 979-8-88697-071-5 (eBook)

Earthworm Engineering and Applications
Adarsh Pal Vig, PhD (Editor)
Surendra Singh Suthar, PhD (Editor)
Jaswinder Singh, PhD (Editor)
2022. ISBN: 978-1-68507-566-8 (Hardcover)
2022. ISBN: 978-1-68507-641-2 (eBook)

Earthworms and their Ecological Significance
Adarsh Pal Vig, PhD (Editor)
Surendra Singh Suthar, PhD (Editor)
Jaswinder Singh, PhD (Editor)
2022. ISBN: 978-1-68507-567-5 (Hardcover)
2022. ISBN: 978-1-68507-632-0 (eBook)

The World of Molecular Biology
Ann F. Varela, Manuel F. Varela, PhD, Michael Shaughnessy (Author)
2021. ISBN: 978-1-53619-232-2 (Hardcover)
2021. ISBN: 978-1-53619-312-1 (eBook)

More information about this series can be found at
https://novapublishers.com/product-category/series/life-sciences-research-and-development/

Michael Y. Wilkerson
Editor

Properties and Applications of Alginate

Copyright © 2022 by Nova Science Publishers, Inc.

All rights reserved. No part of this book may be reproduced, stored in a retrieval system or transmitted in any form or by any means: electronic, electrostatic, magnetic, tape, mechanical photocopying, recording or otherwise without the written permission of the Publisher.

We have partnered with Copyright Clearance Center to make it easy for you to obtain permissions to reuse content from this publication. Simply navigate to this publication's page on Nova's website and locate the "Get Permission" button below the title description. This button is linked directly to the title's permission page on copyright.com. Alternatively, you can visit copyright.com and search by title, ISBN, or ISSN.

For further questions about using the service on copyright.com, please contact:
Copyright Clearance Center
Phone: +1-(978) 750-8400   Fax: +1-(978) 750-4470   E-mail: info@copyright.com

NOTICE TO THE READER

The Publisher has taken reasonable care in the preparation of this book, but makes no expressed or implied warranty of any kind and assumes no responsibility for any errors or omissions. No liability is assumed for incidental or consequential damages in connection with or arising out of information contained in this book. The Publisher shall not be liable for any special, consequential, or exemplary damages resulting, in whole or in part, from the readers' use of, or reliance upon, this material. Any parts of this book based on government reports are so indicated and copyright is claimed for those parts to the extent applicable to compilations of such works.

Independent verification should be sought for any data, advice or recommendations contained in this book. In addition, no responsibility is assumed by the Publisher for any injury and/or damage to persons or property arising from any methods, products, instructions, ideas or otherwise contained in this publication.

This publication is designed to provide accurate and authoritative information with regard to the subject matter covered herein. It is sold with the clear understanding that the Publisher is not engaged in rendering legal or any other professional services. If legal or any other expert assistance is required, the services of a competent person should be sought. FROM A DECLARATION OF PARTICIPANTS JOINTLY ADOPTED BY A COMMITTEE OF THE AMERICAN BAR ASSOCIATION AND A COMMITTEE OF PUBLISHERS.

Additional color graphics may be available in the e-book version of this book.

Library of Congress Cataloging-in-Publication Data

ISBN: 979-8-88697-371-6

Published by Nova Science Publishers, Inc. † New York

Contents

Preface		vii
Chapter 1	Alginate Properties and Applications	1
	Wei Liu, Hui-Jing Li and Yan-Chao Wu	
Chapter 2	Submicron and Nano-Sized Gel Particles Based on Alginate and Sulfated Alginate for Protein Encapsulation	41
	A. Anitha, Jiankun Yang, Yingjun Gao, Auriane Gueguen, Broden Diggle and Lisbeth Grøndahl	
Chapter 3	An Insight into Properties and Applications of Alginate	75
	Abhishek Saxena and Archana Tiwari	
Chapter 4	Alginate Hydrogel Beads as Building Blocks for Delivering Polyphenols	103
	Ina Ćorković, Drago Šubarić, Jurislav Babić, Anita Pichler, Josip Šimunović and Mirela Kopjar	
Chapter 5	Alginate as a Carrier for Probiotic Immobilization: Extrusion and Spray Dry Technique	131
	Tanja Krunic	
Chapter 6	Structure-Property Relationship of Alginate Polymers and Its Biomedical Applications	151
	Bishnu Dev Patra, Sweta Behera, Smrutirekha Mishra, Sankha Chakrabortty and Shirsendu Banerjee	

Chapter 7	**Properties and Medical Applications of Alginate** ....165 Seyed Rasoul Tahami and Nahid Hassanzadeh Nemati
Index	179

Preface

This book includes seven chapters, each focusing on modern research into properties and applications of alginate. The first chapter elucidates the reader to the structural properties and applications of alginate, and additionally discusses future development targets in various fields. The second chapter gives an overview into fabrication of ALG- and S-ALG-based submicron- and nano-sized gel particles with a focus on protein encapsulation. The third chapter returns to analyzing the unique physical properties of alginate and it's current and potential uses. The fourth chapter reviews the development of alginate hydrogel beads by the process of microencapsulation enables the preservation of the entrapped material. The fifth chapter reviews new uses of alginate in the probiotic food industry as a carrier for probiotic immobilization, and summarizes the extrusion and spray dry techniques used to implement them. The sixth chapter gives a brief knowledge of the structure-property relationships, film formation, biocompatibility, and toxicity study of alginate and its different biomedical applications. The seventh and final chapter, after reviewing how alginate is discovered, describes the main properties of alginate, discusses the sources and chemical structure of alginate, and its various applications, especially in drug release and tissue engineering.

Chapter 1 - Alginates are linear biopolymers of 1,4-linked β-D-mannuronic acid (M) and 1,4 α-L-guluronic acid (G) residues, which are arranged in homogenous (poly-G, poly-M) or heterogenous (MG) block-like patterns. The physiological and chemical characteristics of alginates depend on this arrangement of residues. Alginates are very abundant in nature. In brown algae, they are produced as the skeletal component of their cell walls, comprising up to 40% of dry weight. Some bacteria can also synthesize alginates. Alginates have excellent biocompatibility, biodegradation, stable structure, and nontoxicity. Accordingly, they are widely used in biomedicine (such as shaped tablets, dentistry, and wound dressing), food industry (such as ice cream, jelly, lactic beverage, dressing, instant noodle, and beer), animal feed, textile printing industry, papermaking, cosmetic industry, sewage treatment, and daily chemicals. As attractive compounds, alginates have attracted increasing attention in the research world. This chapter summarizes their structures, properties, the modification of alginates, various characterization techniques, processing methods, and dynamic properties. The wide applications of alginates in biomedicine, food, industrial and environmental sciences are reviewed. The future development targets in various fields are prospected.

Chapter 2 - Alginate (ALG) and its heparin (HEP) mimetic derivative sulfated alginate (S-ALG) have been widely studied for use in various biomedical applications. This chapter gives an overview into fabrication of ALG- and S-ALG-based submicron- and nano-sized gel particles with a focus on protein encapsulation. Included is the authors' research on the optimisation of gel particles composed of ALG or S-ALG for the encapsulation of the high pI protein Lactoferrin (Lf). The binding of Lf to ALG and S-ALG as studied through surface plasmon resonance (SPR) revealed non-specific binding to ALG while for S-ALG an association constant, $K_A = 2.0 \times 10^7$ M^{-1}, was determined. The gel particles were synthesised using a nanoprecipitation method and crosslinked with chitosan (CHI) and calcium ions. The particle size characterised using dynamic light scattering (DLS) and nanoparticle tracking analysis (NTA) was in the range of 200–600 nm depending on the reaction conditions. Encapsulation of Lf reduced the particle size for all particles and calcium ions for ionic cross linking was only required for Lf encapsulation in ALG-based gel particles. Suspension stability was evaluated through DLS measurements revealing that only particles produced using nanoprecipitation and sonication had low polydispersity over a 6-day period.

Chapter 3 - Alginates, which are natural multifunctional polymers, have gained popularity in the biomedical and pharmaceutical industries in recent decades due to their unusual physicochemical features and diverse biological activities. Film-forming ability, pH responsiveness, gelling, hydrophilicity, biocompatibility, biodegradability, non-toxic, processability, and ionic crosslinking are only a few of the features of alginates. Food, pharmaceuticals, dental uses, welding rods, and scaffolding are just a few of the commercial applications of alginate. The cosmetics and healthcare sectors have demonstrated a significant deal of interest in biodegradable polymers in general, and alginates in particular, during the last few decades due to their gelling and non-toxic qualities, as well as their abundance in nature. The purpose of this chapter is to describe alginates' unique properties, as well as to examine their current and potential applications.

Chapter 4 - Alginate is a biopolymer which, due to its biocompatible and biodegradable properties has numerous potential applications in various food industry sectors. Development of alginate hydrogel beads by the process of microencapsulation enables the preservation of the entrapped material. As encapsulated materials polyphenols were selected. Polyphenols are well-known health-promoting components which are increasingly explored. Through the improvement of their stability, their range of applications is expanding. The purpose of this chapter was to build upon the available literature data concerning polyphenols as encapsulated materials in alginate hydrogel beads. The main challenges concerning the alginate porous structure were addressed as well. Considering the versatile properties of polyphenols and alginate, these components certainly deserve the attention of the scientific community in order to be fully exploited.

Chapter 5 - Alginate is a natural polysaccharide, which can form a gel and has simple preparation conditions and processes. Also, alginate is a widely used carrier in the food industry due to its pH responsivity, which is important for gastrointestinal digestion. It has the characteristics of biosafety, and biocompatibility and has been widely used and studied in recent years. Alginate is used for probiotic immobilization by extrusion technique due to its ability to form a gel in mild conditions with satisficed the mechanical and chemical stability of carriers. The paper aims to summarize different uses of alginate as a carrier by extrusion and spray dry technique. The viability of probiotic cells is very important due to their numerous health benefits. The desired number of viable bacteria is difficult to achieve as their number decreases due to many environmental factors. Probiotic immobilization increases the viability of bacteria against unfavorable environmental conditions during processing, storage, and gastric and intestinal digestion (conditions such as low pH, presence of digestive enzymes, and bile salt in a human gastrointestinal tract). Different immobilization techniques and the addition of many agents to alginate carriers create beads in a wide range of diameters, with changed fermentative activity and viability of probiotics. This paper reviews the application of sodium alginate in the fields of the probiotic food industry.

Chapter 6 - Alginate is a well-known biomaterial-based polymer composed mainly of polysaccharides. It comes under the category of a biopolymer having α-L-Guluronic acid and β-D-Mannuronic groups in its backbone. These possess ionic properties due to which they are known as anionic polymers enabling better ionic cross-linking structures. It also has some interesting properties such as pH responsiveness, film-forming nature, biocompatible, biodegradable, and non-toxic in nature which makes it suitable for different biomedical applications. The common biomedical applications are dental implantations, scaffolding, and drug delivery hydrogels. Mostly, due to its gelling and biocompatibility nature, it renders a great interest in biomedical applications recently. Considering its biomedical applications, it highly necessitates understanding its structure-property relationships. Hence, this chapter will give a brief knowledge of the structure-property relationships, film formation, biocompatibility, and toxicity study of alginate and its different biomedical applications.

Chapter 7 - Alginate, which refers to all derivatives of Alginic acid, is a natural anionic exopolysaccharide commonly obtained from brown seaweed or certain species of bacteria called Pseudomonas and Azotobacter. The structure of alginate is heteropolymeric, i.e., it consists of two types of Uronic acid, including β-D-Mannuronic acid (M) and α-L-Guluronic acid (G), which are linked by 1,4-glycosidic bonds. The number and length of M blocks and G blocks and MG blocks in the polymer structure vary significantly between alginates and affect their physical and chemical properties. The most important feature of alginates that makes it possible to use them in various applications, especially biomedical applications, is the formation of hydrogels, which in turn is dependent on divalent cations and their bonding and the formation of crosslinking structures. Due to their unique physicochemical properties, they are widely used in tissue engineering, drug delivery systems, advanced wound dressings, bioadhesives, cosmetics, food, and agriculture, the range of their use is increasing and expanding day by day. This chapter, after reviewing how alginate is discovered, describes the main properties of alginate, discusses the sources and chemical structure of alginate, and its various applications, especially in drug release and tissue engineering.

Chapter 1

Alginate Properties and Applications

Wei Liu, PhD, Hui-Jing Li[*], PhD and Yan-Chao Wu, PhD

Weihai Marine Organism and Medical Technology Research Institute, Harbin Institute of Technology, Weihai, P. R. China

Abstract

Alginates are linear biopolymers of 1,4-linked β-D-mannuronic acid (M) and 1,4 α-L-guluronic acid (G) residues, which are arranged in homogenous (poly-G, poly-M) or heterogenous (MG) block-like patterns. The physiological and chemical characteristics of alginates depend on this arrangement of residues. Alginates are very abundant in nature. In brown algae, they are produced as the skeletal component of their cell walls, comprising up to 40% of dry weight. Some bacteria can also synthesize alginates. Alginates have excellent biocompatibility, biodegradation, stable structure, and nontoxicity. Accordingly, they are widely used in biomedicine (such as shaped tablets, dentistry, and wound dressing), food industry (such as ice cream, jelly, lactic beverage, dressing, instant noodle, and beer), animal feed, textile printing industry, papermaking, cosmetic industry, sewage treatment, and daily chemicals. As attractive compounds, alginates have attracted increasing attention in the research world. This chapter summarizes their structures, properties, the modification of alginates, various characterization techniques, processing methods, and dynamic properties. The wide applications of alginates in biomedicine, food, industrial and environmental sciences are reviewed. The future development targets in various fields are prospected.

---

[*] Corresponding Author's Email: lihj@hit.edu.cn.

In: Properties and Applications of Alginate
Editor: Michael Y. Wilkerson
ISBN: 979-8-88697-371-6
© 2022 Nova Science Publishers, Inc.

Keywords: alginates, polysaccharide, properties, applications

Introduction

Alginates are naturally occurring anionic polysaccharides, which are present as structural components in the cell walls of brown algae and can be produced in bacterial strains such as *Azotobacter* and *Pseudomonas*. The industrial applications of alginates were mainly linked to their water-retaining, gelling, viscosifying, and stabilizing properties. Owing to the mature industrial sectors in many developed countries, the market for alginates and their derivatives is observed to be growing at a modest rate. Alginates possess unique properties, which are harnessed for various applications such as in foods, cosmetics, and fabric products as well as pharmaceutical/biomedical and other industrial purposes. For various reasons, researchers are exploring the possible applications of modified forms of alginates with various structures, functions, and properties, which are synthesized via chemical and physical reactions.

1. Chemical Structure

Alginate polymers are a family of linear unbranched polysaccharides made up of 1,4 α-L-guluronic acid (G) and 1,4-linked β-D-mannuronic acid (M) residues organized in homogenous (poly-G, poly-M) or heterogenous (GM, MG) block patterns (Figure 1). The G and M block patterns and sequences may be different in commercial alginates depending on the source of seaweed used, harvesting season, and geographical location of the seaweed (Knudsen et al., 2017). The random sequence of G and M block chains are composed of regions of alternating GM blocks whose monad, diad, and triad frequencies are determined. Hence, the physicochemical properties and degree of polymerization of the alginate depend on the arrangement of these blocks (Kupper et al., 2001). Alginates possess all four glycosidic linkages (Gombotz and Wee, 1998; Kupper et al., 2001) in their structure: (1) diequatorial linkage in MM block; (2) diaxial linkage in GG block; (3) equatorial axial linkage in MG block; (4) axial equatorial linkage in GM block. There are different derivatives of alginates such as sodium alginate, calcium alginate, and propylene glycol alginate, which possess different properties (Orive et al., 2002). In addition, the rigid six-membered sugar rings and the restricted rotation around the glycosidic linkage make alginate molecules stiff. It increases following the order of MG < MM < GG. Therefore, G-rich alginates generally form hard and brittle gels, while soft and elastic gels are produced by M-rich samples.

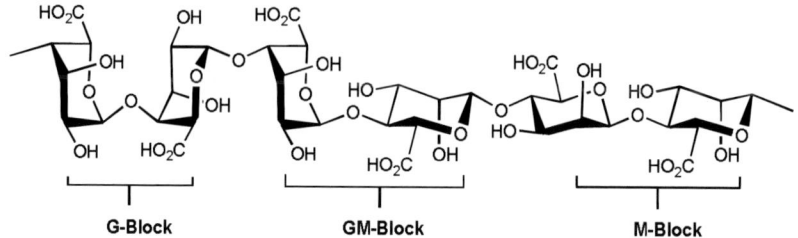

Figure 1. Structure of alginate showing all four glycosidic linkages (GM, MM, MG, and GG block).

2. Sources of Alginates

2.1. Algal Alginates

Alginate is one of the chief products extracted from seaweeds. Alginates of commercial purposes are usually obtained from *Ascophyllum nodosum, Laminaria hyperborea, Macrocystis pyrifera, Durvillaea antarctica, Ecklonia maxima, Lessonia* sp., *Sargassum* sp., and *Turbinaria* sp. These polysaccharides constitute the structural composition of the cell walls and the intercellular matrix in seaweeds. Alginates form 40% of the dry matter of the commercially harvested seaweed species, such as *Laminaria* spp. and *Macrocystis* spp. Various commercial types or forms of alginates are made from brown seaweeds and usually include three steps: pre-extraction, neutralization, and precipitation/purification. Specifically, sodium carbonate (Na_2CO_3), potassium carbonate (K_2CO_3), ammonium carbonate [$(NH_4)_2CO_3$], magnesium carbonate ($MgCO_3$), calcium carbonate ($CaCO_3$) or propylene oxide (C_3H_6O) is added in order to obtain the following alginates, respectively: sodium alginate (Na-alginate), potassium alginate (K-alginate), ammonium alginate (NH_4-alginate), magnesium alginate (Mg-alginate), calcium alginate (Ca-alginate) and propylene glycol alginate (PGA). The extraction procedure of sodium alginate (Critchley, Ohno, and Largo, 2006; Westermeier et al., 2012) was shown in Figure 2.

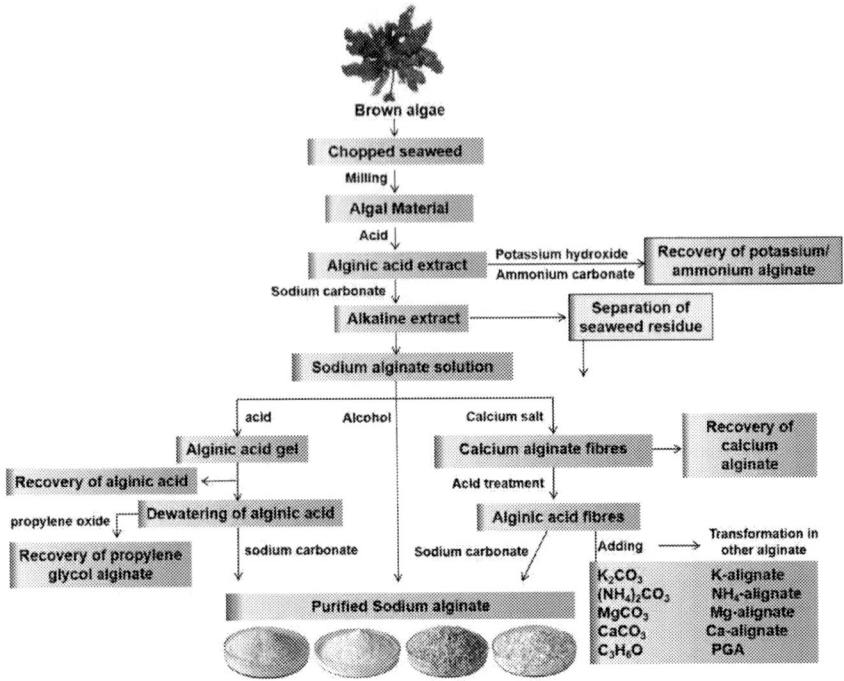

Figure 2. Schematic representation of alginates extraction from brown seaweed.

2.2. Bacterial Alginates

Although brown algae are abundant in the environment, it has several disadvantages such as high production costs, high price, seasonal changes, and other environmental impacts associated with the utilization of these brown seaweeds. These reasons prompted researchers to look for different sources to produce alginic acid. Two bacterial genera, *Pseudomonas* and *Azotobacter*, can produce alginates. Unlike alginates present in brown macroalgae, bacterial-derived polymers are often O-acetylated at O-2 and/or O-3 of D-mannuronate, which is catalysed by mannuronate acetylase (Baker et al., 2014). In nature, microbes produce alginates with various functions through different metabolic processes. The pathway shown in Figure 3 is mainly understood in the bacteria *Pseudomonas aeruginosa* and *Azotobacter vinelandii*, and some homologous genes which have been hypothetically reported in brown alga *Ectocarpus siliculosus* are presented. Gene clusters encoding different proteins accomplishing different steps of alginate biosynthesis are presented and functionally assigned in the frame (the tricarboxylic acid cycle, mannose-6-phosphate isomerase, phosphomannomutase, mannose-1-phosphate guanylyltransferase, Guanosine diphosphate (GDP)-mannose 6-dehydrogenase, mannuronate C5-epimerase, poly-mannuronate, poly-mannuronate/guluronate). The starting point for alginate biosynthesis is fructose 6-phosphate. To date, at least 24 genes have been found to be directly involved in the alginate production in *P. aeruginosa*. Alginate production is mainly considered as a survival advantage for bacteria to survive in unfavorable and harsh conditions (Sabra and Zeng, 2009). *P. aeruginosa* produces alginates that constitute the thick highly structured biofilm, being the characteristic of the species (Nagarajan, Shanmugam, and Zackaria, 2016). Alternatively, *Azotobacter* produces rigid alginate, which is essential for the formation of water-conserving cysts resistant to dessication and stress.

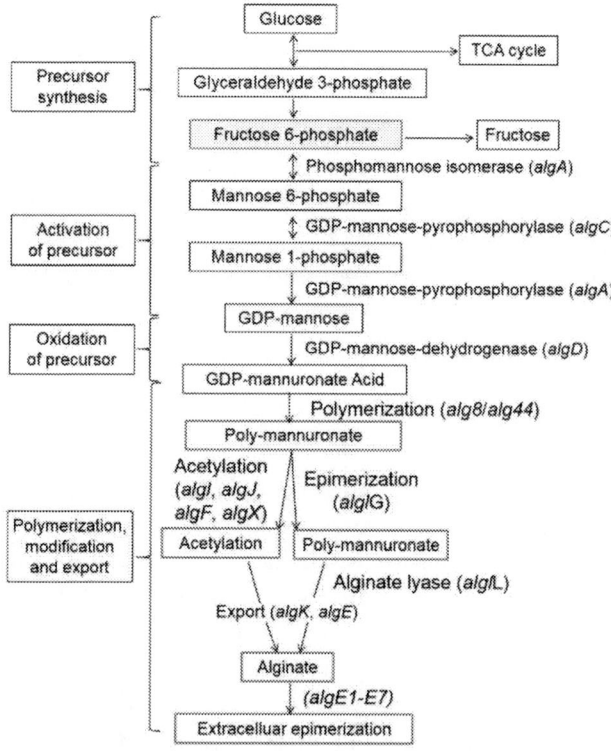

Figure 3. Biosynthesis pathway of alginates in bacteria and algae.

Several other bacterial species, such as *P. mendonica*, *P. putida*, *P. phaseolicola*, *P. savastanoi*, *P. glycinae* and *P. fluorescens,* were also found to produce alginic acid (Govan, Fyfe, and Jarman, 1981; Chen et al., 1985; Fett et al., 1986; Osman, Fett, and Fishman, 1986; Han et al., 2004). Yan et al., (2013) found a bifunctional alginate lyase-producing bacterium strain, which can be used to prepare trisaccharides from alginate. Alginic acid production by bacterial fermentation is not limited by geographical environment and climatic conditions. It can be controlled and optimized by factory scale control production, and the ratio of M/G can also be changed by the change of fermentation conditions to control and optimize production. In recent years, the production of alginates by engineering bacteria has attracted a wide attention (Zhang et al., 2021).

3. Properties of Alginate

The composition, G/M ratio and molecular weight of alginates are known to have effects on their properties. The variability of alginates available with respect to molecular weight, composition, and sequence of G-block and M-block in their copolymer chain changes the viscosity property, capacity to uptake and absorb water, swelling response, and features of sol/gel transition. The chemical and physical properties of alginate are determined by its different components, structure, and conformation. Its chemical and physical properties determine its biological properties.

3.1. Molecular Weight

The molecular weights of alginates generally fall in the range of 32–400 kDa depending on their sources, specieses or extraction processes, and are affected by the G or M blocks present in the alginates. The chemical composition and sequence of the G/M block are affected by the taxonomy, species, age, and type of seaweed, harvesting region, sea current, water temperature, and processing method (Kelly and Brown, 2000; Jothisaraswathi, Babu and Rengasamy, 2006; Rioux, Turgeon, and Beaulieu, 2007). Several techniques for identifying the molecular weight of alginates have been used. Majority of the works concentrate on the sedimentation diffusing viscosity, sedimentation intrinsic viscosity determination, and light scattering (Cook and Smith, 1953;

Wedlock, Fasihuddin, and Phillips, 1986). Sperger et al., (2011) showed that it can be determined by solid-state nuclear magnetic resonance (NMR) even in the presence of water molecules. ^{1}H NMR and Fourier transform infrared spectroscopy techniques are the main techniques applied in the examination of alginates' composition and structural patterns (Pereira and Cotas, 2020).

3.2. Solubility

Alginate is soluble in water but insoluble in organic solvents, whereas alginic acid is insoluble in water but soluble in organic solvents. The solubility of alginate is essential for its application in various industries (Orive et al., 2005). The solubility of alginate is mainly affected by four parameters: (1) The distribution of G and M blocks. Alginate with high G block has higher solubility than that with higher M block; (2) The pH value of the solvent. The pH value of the solution affects the alginate by altering its uronic acid units. It has been found that the decreased solubility of alginate during the commercial process of extraction is due to the removal of a compound called ascophyllan. This compound binds with the alginate to make it more soluble even at low pH values (Haug and Larsen, 1963). Under low pH conditions, alginate with high MG content is soluble, but alginate rich in G- and M-blocks are insoluble (Draget, 2009); (3) Ionic strength. The solubility of alginate generally increases with the increase of ionic strength of the system (except for the gel ions), because high ionic strength can contract molecular chains by electrostatic shielding effect, and decrease the chain winding and the viscosity of the system; (4) The presence of gel ions. It is necessary to use an aqueous solvent free of cross-linking ions to enable the dissolution, since alginate could form gels in the presence of gelling cations such as Ca^{2+}, Ba^{2+} and Sr^{2+}. In addition, alginate with protonated carboxylic acid groups could not be fully dispersed into any solvent including aqueous systems (Cattelan et al., 2020).

3.3. Viscosity

Alginates are soluble in to form a stable and viscous solution. The Federal Communications Commission authorizes sodium alginate viscosities to be between 10–5000 cP. The viscosity of the alginate is affected by many factors, such as the proportion of M and G units in the extracted alginate, the temperature, and the acidic media (Husni et al., 2012). The 1% w/v aqueous solution of sodium alginate has a dynamic viscosity of 20–400 mPa·s at 20°C. With the decrease of pH value, the viscosity of alginate solutions increases and reaches around pH 3–3.5 values, which is due to the protonation of carboxylate groups in the alginate backbone and the formation of hydrogen bonds (Leroux et al., 1999). Physical properties of alginate gels can be modified and further improved by increasing the molecular weight of alginate. However, it becomes highly viscous with the increase of the molecular weight, which is often not desirable in further processing. Therefore, manipulation of the molecular weight and its distribution can independently control the viscosity of the solution before gel formation and subsequent gel stiffness. By changing the combination of high and low molecular weight alginate polymers, the elasticity of gels can be significantly increased with the least increase in solution viscosity (Kong, Lee, and Mooney, 2002). The viscosity of sodium alginate varied significantly among different species of brown seaweeds, ranging from 57.6 to 134.4 cP (Rashedy et al., 2021).

3.4. Stability and Degradation

Alginates can remain stable when not exposed to sunlight, and their molecular weights and functions do not change significantly in dry and cool places within several months. However, alginic acid cannot be kept stable for a long time. A particular physical property of alginate hydrogels, which needs to be considered prior to use, is the degradation behaviour, which can be advantageous or disadvantageous depending on the application. The factors affecting the stability of alginate include acidic hydrolysis, alkaline oxidation, bacterial degradation, and enzymatic degradation. These factors affect the stability of alginate by degrading its molecular structure, thereby reducing its molecular weight. When the molecular weight is affected, the function of alginate will also be affected. Therefore, in order to maintain its molecular weight and stability, it must be protected. Several studies demonstrated the stability of alginates in their various forms (impressions, hydrogels, etc.) since they are used in dental casings and wound dressings (Rad, Ghaffari, and Safavi, 2010; Liu et al., 2016). For enzymatic degradation, alginate lyases catalyze the cleavage of the 1-4 glycosidic bond of alginate via the β-elimination mechanism: polyguluronate lyase that degrades G-blocks, polymannuronate lyase that degrades M-blocks and lyases degrading both blocks. However, alginate is not degraded in the mammalian digestive tract since alginate lyases are not present in our organism. Considering this, hydrolytic degradation is a possible route to remove alginate hydrogel incorporated in our body, being particularly significant factors crosslinking density, pH value and hydrophilicity. Alginate hydrogels are degraded over time under physiological conditions, and thus have limited long-term physiological stability. This degradation occurs via dissipation of the divalent cation crosslinkers as a result of ion exchange when exposed to monovalent cations (such as sodium and potassium). Chelation of divalent cations by phosphate ions can also occur, all of which are present in cell culture media and *in vivo* environments (Bajpai and Sharma, 2004).

3.5. Selective Ion Binding

Alginate has a strong affinity for divalent cations, which are known to bind to the blocks of guluronate residues in adjacent chains. The selective binding of certain metals ions (such as Ca^{2+}, Mg^{2+}, Sr^{2+}, and Ba^{2+}) increases significantly with the increasing content of α-L-guluronate residues in the chains (Smith and Senior, 2021). This binding ability of alginate with the divalent or sometimes the monovalent earth metals is because of its configuration in the G blocks. This mechanism of metal binding to G blocks during gel formation was first described by the "egg box model" (Grant et al., 1973), and is widely accepted as the ionotropic gelation mechanism for alginate. This model explains the binding phenomenon based on the ligands present in the G blocks and the spatial interferences of metals ions with the G blocks (Stokke et al., 2000; Atkins, Mackie and Smolko, 2003). It has been found in previous studies that G blocks can bind with metals, so the polymer with more GG blocks tends to have less binding and therefore less gelling ability (Smidsrod and Haug, 1968; Mancini and McHugh, 2000). The results show that MG blocks can also bind with calcium ions, and contribute to gelation by the binding of calcium ions with adjacent MG blocks and GG blocks.

3.6. Gel Formation Ability

Alginates are best known for their property as a gelling agent. When the pH value decreases, alginic acid will form a gel. Alginate gels have been successfully produced with a low pH value of 2.8–4.0 (King, 1983). The gel has weak gel strength, and the formed gel is soft and soluble in alkaline solution. Sodium alginate can react rapidly with divalent metal ions other than magnesium and mercury to form alginate gels. The strength of the gel film formed with calcium chloride is the largest. The properties of the gel formed are preferably different from the M/G value, the concentration of the sodium alginate, the amount of the combined calcium, and the gelation conditions. Since Na-alginates are soluble in both hot and cold waters, they are used as thickeners, emulsifiers, and gel-forming agents in food industry (Draget et al., 2006). When a small amount of Ca^{2+} is added to the alginic acid solution, Ca^{2+} replaces some of the H^+ and Na^+ in the alginate to form a calcium alginate gel. In this three-dimensional network structure, in the middle of the Ca^{2+} image structure, it forms an "egg-box" structure with the G block. The gel formed by calcium alginate is thermally irreversible, which is an obvious advantage of sodium alginate over other colloids (Strudart et al., 2006; Guo et al., 2020). The gelling of alginate is instantaneous once the addition of calcium is started. It has been pointed out previously that alginates with G blocks have better gelling ability when binding with ions. However, as far as its gelling ability for absorption or immobilization is concerned, M block was found to have better gel swelling ability than the G blocks, and the gel swelling ability of G blocks was found to be increased by addition of sodium ions into the fiber (Qin, 2004). In the field of medicine, Ca-treated alginate gels were found to be used in bone-like apatite formations (Hosoya et al., 2004).

3.7. Biocompatibility

Alginate has good compatibility with human body, so it has been widely used in the medical field. A very high level of biocompatibility is essential as the goal of the encapsulation device is to protect the enclosed cellular tissue from the host's immune response. Although alginate biocompatibility has been extensively investigated, there is a disagreement in the literature. Soon-Shiong (1991) observed a cellular overgrown of 90% of the capsules when high-M block alginate was used. In contrast, Clayton found guluronic acids to be associated with more severe cell overgrowth (Clayton et al., 1991). Some research groups have reported that alginates with a high content in M block evoke an inflammatory response by stimulating monocytes to produce cytokines such as interleukin (IL)-1, IL-6, and tumor necrosis factor (Espevik et al., 1993). Furthermore, antibodies against alginates were found when high-M block alginates were transplanted, but not in the case of high-G block alginates (Kulseng et al., 1999). Vos, Haan, and Schifgaarde (1997) have also reported that after transplantation in rats, most high-G block alginate capsules are overgrown by inflammatory cells and adherent to the abdominal organs, whereas intermediate-G block capsules (with higher M block content) are free of any adhesion and floated freely in the peritoneal cavity. The biocompatibility of the microcapsules is highly influenced by the composition and the purity of the alginate type employed, the biocompatibility of the microcapsule microencapsulation technology and by mechanical factors related to the production process (Orive et al., 2002).

4. Modification of Alginates

Alginates have both hydroxyl and carboxyl functional groups, which are the sites for modification, and the presence of these polar moieties makes alginates hydrophilic in nature. Alginates can undergo chemoselective or stereoselective reaction due to the chemical reactivity of their functional groups. It is essential to know completely the exact chemical reaction that happens in the modification process, otherwise it is difficult to achieve controlled modification. The chemical modification can be classified into hydroxyl modification and carbonyl modification. Chemical modification is the preferred method for generating the alginate derivatives with enhanced and new characteristics (Putri et al., 2021). The range of this modification establishes a remarkable potential for tailoring the next generation of alginate-based biomaterial for applications (Pawar and Edgar, 2012).

4.1. Modification of Carboxyl Group

4.1.1. Esterification of Carboxyl Group
The esterification of alginate is generally carried out in non-aqueous systems. This functionalization could be performed on carbonyl or hydroxyl group (Suzuki et al., 2021). Pawar and Edgar (2013) described the use of tetrabutylammonium fluoride-based two component solvent system as a medium for the synthesis of carboxyl-modified alginate esters. Partially and fully esterified benzyl, butyl, ethyl, and methyl alginates were synthesized via reaction with the corresponding alkylhalides. The newly synthesized derivatives were soluble in polar aprotic solvents. The introduction of butyl groups to alginate through esterification forms hydrophilic alginate without losing its gelation ability in the presence of calcium chloride (Broderick et al.,

2006). By using the same functionalization method, a methyl alginate ester was synthesized and used as an excipient for direct compression to produce immediate drug-release tablets (Sanchez-Ballester et al., 2020). Another study created a nanoparticle platform based on oleate alginate ester for curcumin delivery (Raja, Liu, and Huang, 2015). Alginate has activity to formamide, so it can react with methyl oleate, in which alginate ester is formed after 48 h of reaction and the excess formamide can be removed with soxhlet apparatus.

4.1.2. Carbodiimide Coupling

Carbodiimide is a widely known method to modify alginates that converts the carboxylates into ester and amide derivatives. The reaction is carried out in an aqueous system, and this reaction does not need harsh environments, such as high acid or base. It has been possible to achieve the reversal of predominant hydrophobic interactions between the alkyl chains by mixing β-cyclodextrin with hydrophobically modified alginates (Burckbuchler et al., 2006). β-Cyclodextrin is highly hydrophobic in its interior and external entities also showed similar hydrophobicity, so the small size compounds can be encapsulated within the β-cyclodextrin core. Carbodiimide coupling can achieve the grafting of polyglycolic acid onto the alginate backbone (Chen and Shi, 2015). Others have also reported the application of carbodiimide chemistry to synthesize dodecanol amphiphilic alginate as an emulsifier and hydrogels of tyramine-alginate for bioactive scaffolds, as well as the creation of cell-laden bioinks from thiol-ene crosslinked alginate (Yang et al., 2012; Ooi et al., 2018; Schulz et al., 2019).

4.1.3. Covalent Cross-Linking

Alginate can be crosslinked using covalent as crosslinking agent. The covalently cross-linked alginate was obtained by reacting calcium alginate gels with epichlorohydrin in the presence of base catalyst. The calcium ions may be exchanged with sodium ions without gel dissolution. The carboxylate group of the alginate is preserved in this method, and these modified alginates can be used as ion exchange resin. The network of the gels is insoluble at high ionic strength and when exposed to polar solvents. The water absorption capacity of the modified materials is larger than that of the native alginates (Moe et al., 1993). When the covalent cross-linked alginate is additionally exposed to multivalent ion solutions, the gel strength is higher. The sodium alginate is cross-linked with another alginate in the presence of HCl using the bifunctional aldehyde such as glutaraldehyde to form a cross-linked membrane. The glutaraldehyde cross-linked alginates can be used as a superabsorbent material for sanitary napkins and diapers (Kim et al., 2011). The sodium alginate cross-linked membranes are also useful for the separation of isomers (Yang, Xie, and He, 2011). The covalently cross-linked alginate gels can be utilized to encapsulate active compounds in the pharmaceutical field.

4.2. Chemical Modification of Hydroxyl Group

4.2.1. Esterification of Hydroxyl Groups

Esterification of the hydroxyl groups requires an additional catalyst to achieve a more selective reaction (Suzuki et al., 2021). Alginate has been hydrophobically modified with the dodecyl chain at 12% mol/mol saccharide units (Babak et al., 2000). The alginate acetate was synthesized in dimethyl sulfoxide with a tetrabutylammonium-alginate as the substrate (Pawar and Edgar, 2011). Additionally, ester formation can also be explored with an epoxide ring-opening mechanism to prepare hydrophobically modified alginate (Yu et al., 2014). The condensation of sodium alginate with dodecyl glycidyl ether at pH 9.0 value and 80 °C produces a self-assembled amphiphilic alginate derivative. These polymers have been studied as water-insoluble substance carriers (Meng et al., 2015). This could increase clofazimine and Amphotericin B apparent water solubilities and exhibit a sustained-release effect. In a subsequent report, an improved synthetic method has been developed with the addition of sodium dodecyl sulfate as a surfactant to increase polymer production yields (Wu et al., 2017).

4.2.2. Oxidation

The oxidation reaction is a method to increase the reaction group of sodium alginate and achieve rapid degradation. The oxidation creates aldehyde groups on the alginate. The reactivity of aldehydes is different from that of the alcohols and carboxyl groups, which would help to synthesize new alginate derivatives. Sodium alginate is usually oxidized by oxidative ring-opening method. Generally, the oxidative modification of alginate with sodium periodate is performed on the hydroxyl groups at C-2 and C-3 positions of the uronic units that produce dialdehyde functions on the alginate backbone (Gomez, Rinaudo, and Villar, 2007). These aldehyde groups, due to their high reactivity, can interact with hydroxyl groups on the adjacent uronic acid subunits in the same or neighboring polymer chains to form intramolecular and intermolecular hemiacetals, respectively, leading to the reduction of the available aldehyde groups (Höglund, 2015). Tyramine can be attached to oxidized alginate by reductive amination to improve the immobilization of peroxidase enzyme onto Ca-alginate beads (Prodanovic et al., 2015). At the same time, the aldehyde products of sodium alginate have strong degradation ability, and the degradation performance of sodium alginate is improved. The results show that the oxidation degree and molecular weight of solution can be controlled by changing the concentration of solution at a constant molar ratio of sodium iodate to alginate repeat units, which directly affect the final performance of alginate gelatin hydrogel (Emami et al., 2018).

4.2.3. Sulfation

The hydrogen atom on the hydroxyl group of sodium alginate can be replaced by sulfonation reagent and sulfonation reaction occurs. Sulfation involves the formation of a C-O-S bond, which is mainly utilized to synthesize heparin-like molecules. The chemical sulfation of polysaccharide reaction should be under control, otherwise excess sulfation can occur to result in various side effects. In order to avoid oversulfation, quaternary amine groups are attached to the alginates. The decrease of anticoagulant activity is directly proportional to the number of quaternary amine groups attached onto the alginates. Several studies have reported the development of heparin-like polysaccharides by sulfation, mainly using chlorosulfonic acid as the sulfation agent. Recently, one study discussed the microspheres of the sulfated alginate derivative to promote a drug-controlled release behavior in human induced pluripotent stem cell-derived endothelial cells (Munarin, Kabelac, and Coulombe, 2021). Another strategy employed a carbodiimide-H_2SO_4 intermediate or tetrabutylammonium salt in dimethylformamide (Ma et al., 2016; Freeman, Kedem, and Cohen, 2008). A deteriorating effect on the gelation ability is also found in sulfated alginate, resulting in a lower stiffness gel and higher swelling rate compared with unmodified alginate.

4.2.4. Graft Copolymerization

The physicochemical properties of the alginate were changed by grafting with synthetic polymers. Polyacrylonitrile, polymethyl acrylate or polymethylmethacrylate was used with ceric ammonium nitrate as an initiator for the grafting of alginate. Under the action of the initiator, the hydroxyl group of sodium alginate is triggered as an active free radical, and then reacts with the small molecule monomer by radical polymerization, and the small molecule is grafted onto sodium alginate. Free radical graft polymerization has been performed through atom transfer radical polymerization (Praphakar et al., 2017). The method is utilized to synthesize a macrophage-targeted carrier from sodium alginate to be grafted with allylamine in the presence of ammonium persulfate. Polyacrylamide, when grafted onto alginate backbone, showed high molecular weight, and functioned as a better flocculant in the sedimentation of coal, kaolin, and iron ore slurries.

4.2.5. Phosphorylation

Generally, the phosphorylation reaction in alginate occurs in a dipolar aprotic solvent with urea/phosphoric acid as a reagent system (Coleman et al., 2011). Due to the presence of phosphoric acid, the molar mass and molecular weight decreased during reaction. The degree of substitution of the products is up to 0.26 by this method. The blending of phosphorylated alginate and native alginate gives physical gels, which can resist the leaching of calcium ions. Another study developed an injectable hydrogel from sodium alginate O-phosphorylation, which was achieved using a mixture of $H_3PO_4/P_2O_5/Et_3PO_4$/hexanol (Kim et al., 2015). The internal gelation process produced the hydrogel of the resulting phosphorylated alginate calcium complex and sodium alginate.

5. Types and Applications of Alginates

Commercially, alginate is present in its different salt forms. The types of alginates include sodium alginate, smmonium alginate, sotassium alginate, salcium alginate, sagnesium alginate, and sropylene glycol alginate. The widely used and commercially available alginate products are still mainly sodium alginate, calcium alginate and ammonium alginate. According to the composition and relative molecular weight of alginic acid, the performance diversity of alginic acid must be accompanied by the functional diversity of the substance, that is, the versatility. The uses of alginates are based on three main properties. The first is their ability, when dissolved in water, to thicken the resulting solution (more technically described as their ability to increase the viscosity of aqueous solutions). The second is their ability to form gels with the addition of a calcium salt to a solution of sodium alginate in water. The gel is formed by chemical reaction, in which the calcium displaces the sodium from the alginate, and holds the long alginate molecules together. No heating is required here, and the gels do not melt when heated. The third property of alginates is the ability to form sodium alginate or calcium alginate films and calcium alginate fibers.

5.1. Alginates in Biomedical Applications

5.1.1. Alginates Microcapsules for Drug Delivery

Alginates exhibit excellent biocompatibility and biodegradability that can be useful for many applications in the field of biomedicine (Raus, Wan, and Nasaruddin, 2021). To date, many polymers of have been utilized in cell microencapsulation materials (Mazzitelli et al., 2013), including natural materials such as alginate, chitosan, agarose, collagen, cellulose, and synthetic materials. Among them, alginate remains the most widely employed polymer nowadays. Although other biomaterials may become a good alternative in the future, currently there is a consensus that only alginate is qualified for safe use in human body after exhaustively research (Vos et al., 2014). Moreover, the intrinsic properties of alginate make it suitable for the needs of this biotechnology and confer multiple advantages to the system. Thus, alginate represents a promising tool for drug delivery via cell microencapsulation.

5.1.2. Alginates Application in Tissue Engineering

The strong biocompatibility in cell microenvironment and the possibility of treating alginate solution under safe conditions to afford a stable form after polymer gelation via ionic, chemical, or thermal methods make it useful to design different types of devices (i.e., injectable gels, porous scaffolds, micro-/nanoparticles). These devices are attractive for wound healing, cell transplantation, drug delivery, and three-dimensional scaffolds for tissue engineering applications. They mainly work as ionic polymers derived by the presence of divalent cations such as Ca^{2+}, confering the interesting properties to fabricate micro- and nanostructured devices with improved molecular transport, low cytotoxicity, and relatively low-cost production, which are suitable for large applications in tissue engineering and drug delivery (Gombotz and Wee, 1998). Besides, the hydrogel-like behavior of alginates mainly contributes to the scaffold biomimesis, being structurally like the macromolecular-based components in the body. Their unique chemical properties ensure their full compatibility in the biological microenvironment and minimize the inflammatory reactions after their administration into the body (Sakiyama-Elbert and Hubbell, 2001). In the last few years, alginate has been variously processed to fabricate injectable hydrogels, microspheres, microcapsules, sponges, foams, and fibers to be used as cell or molecular carriers suitable for drug delivery systems or tissue engineering (Varghese and Elisseeff, 2006; Guarino et al., 2016). Dudun et al., (2021) investigated alginate biosynthesis using the *A. vinelandii*. The result showed promising prospects in controlling biosynthesis of bacterial alginate with different physicochemical characteristics for various biomedical applications including tissue engineering.

5.1.3. Alginates Utilization in 3D Cell Culture Models

Three-dimensional cell cultures offer a versatile platform for different needs during various phases of drug discovery and development. Studies have confirmed that the results obtained with 3D cell cultures have better *in vivo* correlation owing to better similarity to the *in vivo* 3D matrix (Imamura et al., 2015). Alginate is a unique biocompatible natural polymer of non-animal origin with structural similarity to extracellular matrix (ECM) (Lee and Mooney, 2012). The traditionally used biocompatible scaffolds are largely based on polymers of animal origin like collagen, hyaluronic acid, and gelatin. The use of such scaffolds for 3D culture leads to significant bioactivity and may hamper the outcomes, whereas alginate is an inert biomaterial with negligible bioactivity, thus making it advantageous over the use of traditional animal derived biopolymers. Alginate is a versatile biomaterial with novel advantages like inertness, animal-free product, no intrinsic bioactivity, and feasible chemical modifications for tailored applications. A significant number of studies have been carried out with alginate-based matrices like 3D scaffolds, alginate beads, alginate hydrogels, alginate collagen complexes, and chitosan alginate gelatin mixed scaffolds. These 3D scaffolds have been widely used in anticancer HTS drug screening, understanding drug molecular mechanisms, molecular pathology of various cancers, enrichment of cancer stem cells, 3D bioprinting, and so on (Florczyk et al., 2012; Kang et al., 2021). In conclusion, alginate-based biomaterials can contribute significantly to expand the horizon of 3D culture and may serve to cut down the cost of commercial 3D anticancer screening platforms for wide-scale application of 3D culture model.

5.1.4. Alginate Application for Heart and Cardiovascular Disease

Alginate has demonstrated great utility and potential as a biomaterial for use in cardiovascular diseases, particularly in the applications such as ECM replacement, 3D microenvironment design for functional cardiac tissue formation, stem cell delivery, and controlled release and presentation of multiple combinations of bioactive molecules and regenerative factors. The most attractive features of alginate for these applications include biocompatibility, mild gelation conditions, and simple modifications to prepare alginate derivatives with new properties. The more recent evolution in cardiac application of alginate has now led to new standards in biomaterial application in the cardiac system not only to "fill the gap" in the injured area, but also to act as an interface with the cardiac biological systems as well (Williams, 2009). These applications focus on four major areas: (1) Alginate hydrogels are used as an ECM substitute in heart tissues to promote tissue regeneration due to their structures are similar to that of natural heart ECM; (2) Alginate hydrogels are used as the carrier of cardiac stem cells or adult cardiomyocytes to the injury site to facilitate the regeneration of functional heart tissue; (3) Using alginate as a platform for delivery of growth factors to mimic natural physiology; (4) Alginate gels are used to control drug release. As a drug release carrier, an alginate-based system can be fine-tuned by controlling the cross-linker type and the cross-linking method to control the speed of cardiac medicine release, such as with antihypertensive or antiarrhythmia medications. Moreover, acellular injectable alginate implants for myocardial repair and tissue reconstruction have entered the stage of clinical research. In addition, 3D printing technique has offered a simple and fast method to fabricate personalized cardiovascular stents with enhanced reproducibility and efficacy (Veerubhotla, Lee, and Chi, 2021). Soon, the use of alginate-based materials in cardiovascular diseases is likely to evolve considerably.

5.1.5. Alginates in Wound Dressings

Alginate has been used to prepare different forms of materials for wound dressings, such as hydrogels, films, wafers, foams, nanofibres, and in topical formulations (Aderibigbe and Buyana, 2018). The wound dressings prepared from alginate can absorb excess wound fluid, maintain a physiologically moist environment, and minimize bacterial infections at the wound site. One of the main reasons for selecting alginate dressings is their ability to absorb wound exudate with alginate dressings capable of absorbing 15–20 times their own weight in wound exudate (Thomas, 1992). Further claims for alginate dressings include treating wound related pain, reducing wound microbial contamination, and reducing wound odour and protease absorption (Chrisman, 2010; Sweeney, Miraftab, and Collyer, 2012). Nanofibres prepared from sodium alginate are potential materials for wound dressing. Nanofibres mimic the extracellular matrix, thereby enhancing the proliferation of epithelial cells and the formation of new tissue (Abrigo, McArthur, and Kingshott, 2014). In addition, sodium alginate has been employed in topical formulation for wound healing. Ahmed et al., (2015) prepared topical formulation from chitosan and sodium alginate loaded with fucidin or Aloe vera with vitamin C using carbopol 934p as a gelling agent. The currently limited use of alginate dressings may offer opportunities for a revival of alginate dressings potentially through the manipulation of the sodium and calcium alginate fibre M- and G- group composition along with the introduction of silver and other antimicrobial agents.

5.1.6. Alginates in Metabolic Syndrome

Alginates are primarily used for the management of gastrointestinal tract disorders, but they are of potential use to attenuate the components of the metabolic syndrome including obesity, type 2 diabetes, hypertension, non-alcoholic fatty liver disease and dyslipidaemia (Williams et al., 2004; Chater et al., 2015). As prebiotics, alginates changed the gut microbiome to increase the production of short-chain fatty acids as substrates for bifidobacteria. Alginates inhibited pancreatic lipases and so decreased triacylglycerol breakdown and uptake. Treatment with alginates decreased food intake by inducing satiety and weight loss in patients on a calorie-restricted diet. Both glucose and fatty acid uptake were reduced. In rat models of hypertension, alginates decreased blood pressure (Chen et al., 2010). An alginate-antacid combination is an effective treatment of gastric reflux disease by forming a raft on the gastric contents (Rohof et al., 2013). Alginates are important as drug carriers in microparticles and nanoparticles to increase drug bioavailability, such as drugs for treatment of metabolic syndrome (Tønnesen and Karlsen, 2002). Alginates are also used to protect cells during transplantation from hos immune responses, allowing potential long-term control of some endocrine disorders such as type 1 diabetes and increased thermogenesis by brown adipocytes in obesity (Fujikura, Hosoda, and Nakao, 2013). These versatile biopolymers have many potential applications in the treatment of human diseases.

5.1.7. Alginates Oligomers as Active Pharmaceutical Drugs

Alginate oligomers retain most of the chemical and physical properties of the higher molecular weight commercial alginates, retaining affinity towards monovalent and divalent ions, which is dependent on the chemical composition of the oligomer. The mucus rheology-modifying properties of alginate oligomers have clinical significance for the cystic fibrosis and other chronic respiratory conditions such as chronic obstructive pulmonary disease (Bazett et al., 2015). Surprisingly, it has been shown that alginate oligomers are able to potentiate the activity of conventional antibiotics (i.e., macrolides and β-lactams) against a range of multidrug-resistant (MDR) pathogens, such as *Pseudomonas, Acinetobacter* and *Burkholderia* spp. (Khan et al., 2012). The potential use of the alginate oligomers in the treatment of infections is considerable, particularly in the treatment of addressing MDR bacterial and fungal infections, where there are limited therapeutic options. Moreover, in contrast to antibiotic treatments that induce resistance, long-term studies with alginate oligomers have not shown any development of resistance. This application of alginate oligomer technology is of particular importance in the treatment of biofilm-related infections (Wang et al., 2016), such as chronic venous leg ulcers, diabetic foot ulcers and periodontal disease, where the persistence of biofilm contributes to the chronic inflammation and infection in these patients. The safety profile and chemical characteristics of alginate oligomers facilitate their application, formulation, and delivery for these multiple therapeutic applications.

5.2. Alginate as a Functional Food Ingredient

5.2.1. Gelling, Thickening, Stabilising and Emulsifying Agent

Alginates are available at a wide range of viscosities, offer stability to foodstuffs under both high and low temperatures, and therefore have a wide range of uses as gelling agents. Alginate gelation rates and gel strengths can be controlled by the concentration of Ca^{2+} or H^+ in the solution. As alginates form gels at low temperatures, this is particularly useful in the restructuring of foodstuffs that may become damaged or oxidised under high temperatures (e.g., meat products, fruits, and vegetables).

Alginates are commonly used as thickening agents in jams, marmalades, and fruit sauces as alginate-pectin interactions are thermally reversible and have a higher viscosity than any individual component (Brownlee et al., 2005). PGAs are often used for this application, but at concentrations lower than those of standard alginate salts. Alginates form stable gels at high and low temperatures and at low pH values (Smidsrod and Draget, 1996). As a result, they can be used for several stability applications in food processing (Hubbermann et al., 2006; Paraskevopoulou, Boskou, and Paraskevopoulou, 2006). Routine use of alginates in bakery creams endows the cream with freeze/thaw stability and reduces separation of the solid and liquid components (syneresis). Alginates are used in combination with other hydrocolloids to thicken and stabilise ice cream (Regand and Goff, 2003). PGAs are commonly used to maintain foam stability, including applications in mousse and other desserts (Baeza et al., 2004). The largest food use of PGA is within the brewing industry, where PGA added to different beers and lagers stabilises the froth head when poured, while also protecting it from foam-negative contaminants (Ferreira et al., 2005). Hydrocolloids are water-soluble and very hydrophilic. However, previous studies have suggested that, within their use as stabilisers in food oil-water emulsions, they exhibit the ability to precipitate/adsorb onto oil droplets and sterically stabilise emulsions against flocculation and coalescence. PGAs have previously been demonstrated to be better emulsifiers than methylcellulose compounds and locust bean and guar gums, and to be comparable with other commonly used plant-derived polysaccharides (i.e., pectin and gum arabic) on a weight-for-weight basis (Huang, Kakuda, and Cui, 2001). These factors are useful in simple food emulsions in which PGA and alginates are utilised, such as mayonnaise and other dressings (Mancini et al., 2002; Paraskevopoulou et al., 2005).

5.2.2. Encapsulation and Immobilisation

In the food industry, alginate encapsulation and immobilisation technologies are used for a variety of purposes, including food processing, food functionality and product acceptability. These immobilisation technologies can enhance productivity as a result of continuous operation and reuse of the entrapped cells or enzymes (Groboillot et al., 1994; Liu et al., 2020). Immobilisation or encapsulation technology is used to produce a wide range of bacterial metabolites, including enzymes, amino acids, organic acids (e.g., acetic acid) and alcohols (Norton and Vuillemard, 1994). Alginates are commonly used as an immobilisation medium for whole cell or isolated enzyme preparations in industrial processes requiring enzymatic activity. Immobilisation in this manner is a simple means of reducing cell or enzyme loss, and can improve the heat stability of entrapped enzymes. As alginates (particularly their calcium salts) form ionotropic gels spontaneously under low-temperature conditions, they are ideal for entrapment of whole cells or enzymes, which would otherwise be damaged under more stringent conditions (Navratil et al., 2002). One of the latest developments in alginate use by the food industry is the encapsulation of live cells (probiotics) for delivery to the human large bowel (Kailasapathy, 2006). As alginates form highly porous gels, they may not be suitable for the encapsulation of reactive or volatile molecules, such as acidulants, fats and flavours (Desai and Park, 2005).

5.2.3. Food Coating

Alginates are one of a wide range of polysaccharides and proteins that have been used as an edible coating for a variety of foodstuffs. Food coatings formed from sodium alginate have been shown to have excellent tensile strength, flexibility, and resistance to tearing, and are impermeable to oils. However, owing to the porous nature of alginate gels, these coatings tend to have high permeability to oxygen and water (Wang et al., 2007). Alginate-based edible coatings and films attract interest in improving/maintaining the quality and extending the shelf-life of fruit, vegetable, meat, poultry, seafood, and cheese by reducing dehydration (as sacrificial moisture agent), controlling respiration, enhancing product appearance, improving mechanical properties, etc (Parreidt, Müller, and Schmid, 2018). Alginate food coatings can be formed ionotropically at room temperature, and therefore their use may be advantageous in several food applications. Incorporation of antimicrobial agents in the alginate gel has been demonstrated to be an effective barrier to microbial surface spoilage of vegetables, meat (Oussalah et al., 2007) and fish products (Datta et al., 2008). The cold formation of the alginate gel coating reduces the damage to the antimicrobial entities and the foodstuff itself. This property has also been demonstrated to be useful in the coating of a variety of fresh fruits and vegetable products, such as lettuce (Tay and Perera, 2004) and freshly cut apple and melon slices (Oms-Oliu, Soliva-Fortuny, and Martin-Belloso, 2008), where the coating can extend the shelf life, reduce browning, maintaine crispness and texture, and reduce vitamin C loss.

5.3. Alginate Application in Industrial and Environmental Sciences

5.3.1. Water Treatment of Heavy Metal Ions

Industrial settings often lead to water pollution by bioaccumulating toxic metals that do not undergo natural biodegradation. The biodegradability and abundancy of alginate have attracted attention for their use in heavy metal ions removal from water effluents (Facchi et al., 2018). Dubey, Bajpai, and Bajpai (2016) have prepared chitosan-alginate nanoparticles (NPs) for the effective and economically viable removal of Hg^{2+} ions. Similarly, Almutairi et al., (2021) have studied the removal of hexavalent chromium (Cr^{6+}) from polluted water using negatively charged (-23.2 mV) chitosan-alginate nanocomposites incorporated with iron NPs. In contrast, Ahmed et al., (2015) have developed more complex nanocomposites comprising green nano-zerovalent copper, activated carbon, chitosan, and alginate. These nanosystems were able to remove Cr^{6+} from polluted water in a proportion of up to 97.5%. The alginate could be also used for the selective removal of heavy metals such as lead and nickel (Vold et al., 2003). Córdova et al., (2021) confirmed the great potential of using xanthate-modified alginate to remove Ni(II) ions for further application of metal-containing derivative as flame ratardant additive. Likewise, encouraging results were obtained in the removal of Pb(II) ions, but more analysis is necessary to find suitable applications of metal-containing derivative. The high affinity of alginate towards Pb(II) and Ni(II) indicates that this polysaccharide consists of β-D-mannuronic acid and α-L-guluronic acid, in which the carboxylic groups alongside the polyguluronic chains promote an efficient interaction involving these cations (Papageorgiou et al., 2010).

5.3.2. Textile Industries

The wastewater produced from textile industries contains a lot of waste dyes and pigments, leading to environmental pollution. The colored wastewater from the textile industries is rated as the most polluted wastewater in almost all industrial sectors (Andleeb et al., 2010). Researchers have tried to develop an inexpensive and efficient method for the treatment of disperses dyes that are potentially toxic and even carcinogenic. Due to its gel ability, calcium alginate has been investigated as a possible coagulant. The development and application of horseradish peroxidase immobilized on calcium alginate gel beads has been reported for successful and effective decolorization of textile industrial effluent (Gholami-Borujeni et al., 2011). The effective removal of methylene blue from aqueous solution in a batch stirred tank reactor was done by using alginate/polyvinyl alcohol–kaolin clay composite material by El-Latif et al., (2010). Boucherit, Abouseoud, and Adoura (2012) reported the decolorization of disperse dye in free and immobilized forms by calcium alginate gel entrapment process using cucurbita peroxidase extracted from courgetti. Calcium alginate, as a green carrier for TiO_2 immobilization, can be used for developing a new environmentally friendly immobilization system for large-scale water treatment. Polyitaconic acid grafted sodium alginate, NaAlg/IA, was employed in studies on the adsorption kinetic of Pb^{2+} in aqueous solution (Mahmoud and Mohamed, 2012). It has been validated that the alginate can be used as an effective coagulant for reactive magenta dye removal from textile wastewater in real time (Vijayaraghavan and Shanthakumar, 2018).

5.3.3. Paper Making Industries

Zeng et al., (2011) reported the use of the composite composed of sodium alginate, polyaluminum ferric chloride, and cationic polyacrylamide as the flocculants in treating wastewater from papermaking industries. The decolorization of paper mill effluent by sodium alginate-immobilized cells of *Phanerochaete chrysosporium* was investigated by Gomathi et al., (2012). The results indicate that the combination of calcium alginate-immobilized *P. chrysosporium* with the addition of carbon and nitrogen sources was helpful to achieve high decolorization efficiency. When the immobilized culture was used under aerobic conditions, it was found that the biological oxygen demand and chemical oxygen demand (COD) and other characteristics of the effluent were reduced, and high transparency of the effluent was obtained. In addition, anaerobic beads with 0.85–1.45 diameters were produced by directly injecting and mixing anaerobic sludge and 1.5% sodium alginate solution. The sequencing batch reactor with anaerobic biobeads was used to treat paper wastewater, and about 85% COD removal could be achieved after 6 days (Jung, Han, and Chung, 2018).

5.3.4. Tanneries

Tanning alters the physiochemical property of collagen in animal skins or hides by using agents, yielding functional leather. Tannery waste is a complex mixture that makes the design of effluent treatment challenging. Wattle extract (vegetable-based tanning dye), chrome tannin (residual tanning broth containing chromium), and chemical dye compounds were found to be the three main pollutants in the tannery effluent (Kanagaraj and Mandal, 2012). Thus far, the most dominant tanning agent remains chromium (III) salt. By utilizing sodium alginate, the chemical coagulants such as sodium citrate, ammonium aluminum sulfate, aluminum sulfate, and calcium carbonate were immobilized in the bead form to treat tannery wastewater samples. After immobilization, chemical coagulants were found to be more effective than their natural forms in reducing electrical conductivity, total dissolved solids, COD, and chromium amounts. Biomass/polymer matrices beads (BPMB) prepared by the immobilization of Baker's yeast strain (*Saccharomyces cerevisiae*) biomass in 3% alginate extract were investigated for chromium biosorption from aqueous solution. The results reveal that the prepared BPMB beads offer excellent potential for chromium removal from contaminated sites. Compared with the free cells, the cells of marine microalgae *Chlorella salina* immobilized with sodium alginate are a promising biological remediator to remove phosphate, nitrate, and ammonia from the tannery industrial effluent, and improve the water quality to its discharging limits. The immobilized bacterium *P. putida* can be used to effectively treat various industrial effluents (Gidhamaari, Boominathan, and Mamidala, 2012). Guan et al., (2022) proposed a strategy integrating chrchromium (III)-containing tanning effluent treatment and retanning/filling resource utilization of adsorbent/adsorbate, which is an effective alternative on account of its convenient operation, environmental sustainability, and wide adaptability.

5.3.5. Release Agents

Alginate films have poor adhesion to many surfaces and are insoluble in nonaqueous solvents, resulting in their initial use as mould release agents for plaster moulds (Andreopoulos, 1987) and later for the forming of fibreglass plastics (Bogun, 2010). Sodium alginate can be used to make a good coating for anti-tack paper and as a release agent in the manufacture of synthetic resin decorative boards. Films of calcium alginate, formed in situ on a paper, have been used to separate decorative laminates after they have been formed in a hot-pressing system.

5.4. Agricultural Applications

Alginate is also used in agriculture as an additive to fertilizers or to make fertilizers with alginic acid. The fertilizer has the advantages of incomparable natural property, non-toxic and side effects, and has a very broad application prospect. In addition, the possible use as a soil conditioner of Na-alginate copolymers was investigated by Hassan and El-Rehim (2006). The addition of polyacrylamide/sodium alginate/Na-alginate copolymer in small quantities to sandy soil increased its ability to retain water and to promote plant growth. Another interesting use of alginate in agriculture is to encapsulate insecticides to reduce their adverse effect on the environment. Kaur, Agnihotri, and Goyal (2021) loaded cartap hydrochloride into chitosan-alginate nanosphere using ionic gel and polyelectrolyte complexation technology. As the encapsulation efficiency is higher than 75% and the stability can be maintained for 30 days at ambient temperature, the developed nanosystem is considered promising for reducing field application frequency. Moreover, its slow release to the target organism is also economically relevant and safe for the environment. Alternatively, Kumar et al., (2015) fabricated alginate-chitosan-based nanocapsules loaded with acetamiprid. The as-described nanoformulation has the potential to reduce the application frequency of pesticides by controlling the release of agrochemicals to reduce the associated side effects. Herbicides can also be loaded into chitosan-alginate nanoparticles.

Future Direction for Research

Alginates have been developed in the field of pharmaceutics and food industries due to their flexible physical, chemical, and physiological properties. Alginates have been used in drugs, protein delivery, tissue regeneration, wound dressing, and wound healing in pharmaceutical industry. In food industry, alginates have been used as emulsifiers, stabilizers, texturizers, etc., and in paper industry to improve the crumpling and resistance of paper. Alginates have been also used in textile, cosmetics, and welding industries as thickeners and as a binder of flux, respectively. These advantageous characteristics and the wide use of alginate make them a promising candidate for applications. In the future, efforts should be made to improve the alginate biomaterial and further improve their application effect. The liquid handling capacity of sodium alginate dressing may be improved and has other advantages such as antibacterial ability. Alginates and their derivatives have great potentials in the treatment of wastewater. Moreover, new biosorbents should be developed by modification of alginate with various other materials to explore their innovative performances against environmental pollution.

References

Abrigo, M., McArthur, S. L. and Kingshott, P. (2014). Electrospun nanofibers as dressings for chronic wound care: Advances, challenges, and future prospects. *Macromolecular Bioscience*, 772-792.
Aderibigbe, B. and Buyana, B. (2018). Alginate in wound dressings. *Pharmaceutics*, 42.
Ahmed, M. M., Jahangir, M. A., Saleem, M. A., Kazmi, I., Bhavani, P. D. and Muheem A. (2015). Formulation and evaluation of fucidin topical gel containing wound healing modifiers. *American Journal of PharmTech Research*, 232-242.
Almutairi, F. M., El Rabey, H. A., Alalawy, A. I., Salama, A. A. M., Tayel, A. A., Mohammed, G. M., Aljohani, M. M., Keshk, A. A., Abbas, N. H. and Zayed, M. M. (2021). Application of chitosan/alginate nanocomposite incorporated with phycosynthesized iron nanoparticles for efficient remediation of chromium. *Polymers*, 2481.

Andleeb, S., Atiq, N., Ali, M. I., Razi-UL-Hussain, R., Shafique, M., Ahmed, B., Ghumro, P. B., Hussain, M., Hameed, A. and Ahmed. S. (2010). Biological treatment of textile effluent in stirred tank bioreactor. *International Journal of Agriculture and Biology*, 256-260.

Andreopoulos, A. G. (1987). Diffusion characteristics of alginate membranes. *Biomaterials*, 397-400.

Atkins, E. D. T., Mackie, W. and Smolko, E. E. (1970). Crystalline structures of alginic acids. *Nature*, 626-628.

Babak, V. G., Skotnikova, E. A., Lukina, I. G., Pelletier, S., Hubert, P. and Dellacherie, E. (2000). Hydrophobically associating alginate derivatives: Surface tension properties of their mixed aqueous solutions with oppositely charged surfactants. *Journal of Colloid and Interface Science*, 505-510.

Baeza, R., Sanchez, C. C., Pilosof, A. M. R. and Patino, J. M. R. (2004). Interfacial and foaming properties of prolylenglycol alginates: effect of degree of esterification and molecular weight. *Colloids Surf B Biointerfaces*, 139-145.

Bajpai, S. K. and Sharma, S. (2004). Investigation of swelling/degradation behaviour of alginate beads crosslinked with Ca^{2+} and Ba^{2+} ions. *Reactive and Functional Polymers*, 129-140.

Baker, P., Ricer, T., Moynihan, P. J., Kitova, E. N., Walvoort, M. T. C., Little, D. J., Whitne, J. C., Dawson, K., Weadge, J. T., Robinson, H., Ohman, D. E., Codée, J. D. C., Klassen, J. S., Clarke, A. J. and Howell, P. L. (2014). *P. aeruginosa* SGNH hydrolase-like proteins AlgJ and AlgX have similar topology but separate and distinct roles in alginate acetylation. *Plos Pathogens*. e1004334.

Bazett, M., Honeyman, L., Stefanov, A. N., Pope, C. E., Hoffman, L. R. and Haston, C. K. (2015). Cystic fibrosis mouse model-dependent intestinal structure and gut microbiome. *Mammalian Genome*, 222-223.

Bogun, M. (2010). Nanocomposite calcium alginate fibres containing SiO_2 and bioglass. *Fibres & Textiles in Eastern Europe*, 11-19.

Boucherit, N., Abouseoud, M. and Adoura, L. (2012). Degradation of disperse dye from textile effluent by free and immobilized peroxidase. *Journal of Environmental Sciences*, 1235-1244.

Broderick, E., Lyons, H., Pembroke, T., Byrne, H., Murray, B. and Hall, M. (2006). The characterisation of a novel, covalently modified, amphiphilic alginate derivative, which retains gelling and non-toxic properties. *Journal of Colloid and Interface Science*, 154-161.

Brownlee, I. A., Allen, A., Pearson, J. P., Dettmar, P. W., Havler, M. E., Atherton, M. R. and Onsoyen, E. (2005). Alginate as a source of dietary fiber. *Critical Reviews in Food Science and Nutrition*, 497-451.

Burckbuchler, V., Kjøniksen, A. L., Galant, C., Lund, R., Amiel, C., Knudsen, K. D. and Nyström, B. (2006). Rheological and structural characterization of the interactions between cyclodextrin compounds and hydrophobically modified alginate. *Biomacromolecules*, 1871-1878.

Cattelan, G., Gerbolés, A. G., Foresti, R., Pramstaller, P. P., Rossini, A., Miragoli, M. and Malvezzi, C. C. (2020). Alginate formulations: Current developments in the race for hydrogel-based cardiac regeneration, *Frontiers in Bioengineering and Biotechnology*, 414.

Chater, P. I., Wilcox, M. D., Houghton, D. and Pearson, J. P. (2015). The role of seaweed bioactives in the control of digestion: implications for obesity treatments. *Food & Function*, 3420-3427.

Chen, M. and Shi, Q. (2015). Transforming sugarcane bagasse into bioplastics via homogeneous modification with phthalic anhydride in ionic liquid. *ACS Sustainable Chemistry & Engineering*, 2510-2515.

Chen, W., Chen, J., Chang, S. and Su, C. (1985). Bacterial alginate produced by a mutant of *Azotobacter vinelandii*. *Applied and Environmental Microbiology*, 543-546.

Chen, Y., Ji, W., Du, J., Yu, D., He, Y., Yu, C., Li, D., Zhao, C. and Qiao, K. (2010). Preventive effects of low molecular mass potassium alginate extracted from brown algae on DOCA salt-induced hypertension in rats. *Biomed Pharmacother*, 291-229.

Chrisman, C. A. (2010). Care of chronic wounds in palliative care and end-of-life patients. *International Wound Journal*, 214-235.

Clayton, H. A., London, N. J., Colloby, P. S., Bell, P. R. and James, R. F. (1991). The effect of capsule composition on the biocompatibility of alginate-poly-1-lysine capsules. *Journal of Microencapsulation*, 221-233.

Coleman, R. J., Lawrie, G., Lambert, L. K., Whittaker, M., Jack, K. S. and Grøndahl, L. (2011). Phosphorylation of alginate: Synthesis, characterization, and evaluation of *in vitro* mineralization capacity. *Biomacromolecules*, 889-897.

Cook, W. H. and Smith, D. B. (1953). Molecular weight and hydrodynamic properties of sodium alginate. *Canadian Journal of Chemistry*. 227-239.

Córdova, B. M., Infantas, G. C., Mayta, S., Huamani-Palomino, R. G., Kock, F. V. C., Oca, J. M. D. and Valderrama, A. C. (2021). Xanthate-modified alginates for the removal of Pb(II) and Ni(II) from aqueous solutions: A brief analysis of alginate xanthation. *International Journal of Biological Macromolecules*, 557-566.

Critchley, A. T., Ohno, M. and Largo, D. B. (eds.) (2006). *World seaweed resources, in: DVD-ROM, Expert Centre for Taxonomic Identification* (ETI), University of Amsterdam, Amsterdam, Springer, New York.

Datta, S., Janes, M. E., Xue, Q., Losso, J. and La Peyre, J. F. (2008). Control of *Listeria monocytogenes* and *Salmonella anatum* on the surface of smoked salmon coated with calcium alginate coating containing oyster lysozyme and nisin. *Journal of Food Science*, 67-71.

Desai, K. G. H. and Park, H. J. (2005). Recent developments in microencapsulation of food ingredients. *Drying Technology*, 1361-1394.

Draget, K. I. (2009). Alginates, in: *Handbook of hydrocolloids*. Phillips, G. O. and Williams, P. A. (Eds.), Woodhead Publishing, Cambridge, 807-828.

Draget, K. I., Moe, S. T., Skjak-Brek, G. and Smidsrod, O. (2006). Alginate, in: Food polysaccharides and their application. Stephen, A. M., Phillips, G. O. and Williams, P. A. (Eds.), Taylor & Francis Group, LLC, Boca Raton.

Dubey, R., Bajpai, J. and Bajpai, A. K. (2016). Chitosan-alginate nanoparticles (CANPs) as potential nanosorbent for removal of Hg (II) ions. *Environmental Nanotechnology, Monitoring & Management*, 32-44.

Dudun, A., Akoulina, A. E., Zhuikov, A. V., Makhina, T. K., Voinova, V. V., Belishev, N. V., Khaydapova, D. D., Shaitan, K. V., Bonartseva, G. A. and Bonartsev, A. P. (2021). Competitive biosynthesis of bacterial alginate using *Azotobacter vinelandii* 12 for tissue engineering applications. *Polymers*, 131.

El-Latif, M. A. M., El-Kady, M. F., Ibrahim, A. M. and Ossman, M. E. (2010). Alginate/polyvinyl alcohol-Kaolin composite for removal of methylene blue from aqueous solution in a batch stirred tank reactor. *Journal of American Science*, 280-292.

Emami, Z., Ehsani, M., Zandi, M. and Foudazi, Z. (2018). Controlling alginate oxidation conditions for making alginate-gelatin hydrogels. *Carbohydrate Polymers*, 509-517.

Espevik, T., Ottrerlei, M., Skjak-Bræk, G., Ryan, L., Wright, S. D. and Sundan, A. (1993). The involvment of CD14 in stimulation of cytokine production by uronic acid polymers. *European Journal of Immunology*, 255-261.

Facchi, D. P., Cazetta, A. L., Canesin, E. A., Almeida, V. C., Bonafé, E. G., Kipper, M. J. and Martins, A. F. (2018). New magnetic chitosan/alginate/Fe_3O_4@SiO_2 hydrogel composites applied for removal of Pb(II) ions from aqueous systems. *Chemical Engineering Journal*, 595-608.

Ferreira, I. M. P. L. V., Jorge, K., Nogueira, L. C., Silva, F. and Trugo, L. C. (2005). Effects of the combination of hydrophobic polypeptides, iso-alpha acids, and malto-oligosaccharides on beer foam stability. *Journal of Agricultural and Food Chemistry*, 4976-4981.

Fett, W. F., Osman, S. F., Fishman, M. L. and SIEBLES, T. S. (1986). Alginate production by plant-pathogenic *Pseudomonads*. *Applied and Environmental Microbiology*, 466-473.

Florczyk, S. J., Liu, G., Kievit, F. M., Lewis, A. M., Wu, J. and Zhang, M. (2012). 3D porous chitosan-alginate scaffolds: A new matrix for studying prostate cancer cell-lymphocyte interactions *in vitro*. *Advanced Healthcare Materials*, 590-599.

Freeman, I., Kedem, A. and Cohen, S. (2008). The effect of sulfation of alginate hydrogels on the specific binding and controlled release of heparin-binding proteins. *Biomaterials*, 3260-3268.

Fujikura, J., Hosoda, K., Nakao, K. (2013). Cell transplantation therapy for diabetes mellitus: Endocrine pancreas and adipocyte. *Endocrine Journal*, 697-708.

Gholami-Borujeni, F., Mahvi, A. H., Naseri, S., Faramarzi, M. A., Nabizadeh, R. and Alimohammadi, M. (2011). Application of immobilized horseradish peroxidase for removal and detoxification of azo dye from aqueous solution. *Research Journal of Chemistry and Environment*, 217-222.

Gidhamaari, S., Boominathan, and Mamidala, E. (2012). Studies of efficiency of immobilized bacteria in tannery effluent treatment. *Journal of Bio Innovation*, 33-42.

Gomathi, V. C., Ramanathan, B., Sivaramaiah Nallapeta, A., Ramanjaneya, V., Mula, R. and Jayasimha Rayalu, D. (2012). Decolourization of paper mill effluent by immobilized cells of *Phanerochaete chrysosporium*. *International Journal of Plant, Animal and Environmental Sciences*, 141-146.

Gombotz, W. R. and Wee, S. F. (1998). Protein release from alginate matrices. *Advanced Drug Delivery Reviews*, 267-285.

Gomez, C. G., Rinaudo, M. and Villar, M. A. (2007). Oxidation of sodium alginate and characterization of the oxidized derivatives. *Carbohydrate Polymers*, 296-304.

Govan, J. R. W., Fyfe, J. A. M. and Jarman, T. R. (1981). Isolation of alginate-producing mutants of *Pseudomonas fluorescens*, *Pseudomonas putida* and *Pseudomonas mendocina*. *Journal of General and Applied Microbiology*, 217-220.

Grant, G. T., Morris, E. R., Rees, D. A., Smith, P. J. C. and Thom, D. (1973). Biological interactions between polysaccharides and divalent cations: The egg-box model. *FEBS Letters*, 195-198.

Guan, X., Zhang, B., Li, D., He, M., Han, Q. and Chang, J. (2022). Remediation and resource utilization of chromium(III)-containing tannery effluent based on chitosan-sodium alginate hydrogel. *Carbohydrate Polymers*, 119179.

Guarino, V., D'Albore, M., Altobelli, R. Ambrosio, L. (2016). Polymer bioprocessing to fabricate 3D scaffolds for tissue engineering. *International Polymer Processing Journal of the Polymer Processing Society*, 587-597.

Guo, X., Wang, Y., Qin, Y., Shen, P. and Peng, Q. (2020). Structures, properties and application of alginic acid: A review. *International Journal of Biological Macromolecules*, 618-628.

Han, W., Lu, X., Xiao, L. and Yu, W. (2004). Screening of an alginate-producing marine bacterium *Pseudomonas* sp. strain QDA and its characterization. *Journal of Ocean University of China*, 60-64.

Hassan A. Abd El-Rehim. (2006). Characterization and possible agricultural application of polyacrylamide/sodium alginate crosslinked hydrogels prepared by ionizing radiation. *Journal of Applied Polymer Science*, 3572-3580.

Haug, A. and Larsen, B. (1963). The solubility of alginate at low pH. *Acta Chemica Scandinavica*, 1653-1662.

Höglund, E. (2015). *Production of dialdehyde cellulose and periodate regeneration: Towards feasible oxidation processes*.

Hosoya, K., Ohtsuki, C., Kawai, T., Kamitakahara, M., Ogata, S., Miyazaki, T. and Tanihara, M. (2004). A novel covalently cross-linked gel of alginate and silane with the ability to form bone-like apatite. *Wiley Interscience*.

Huang, X., Kakuda, Y. and Cui, W. (2001). Hydrocolloids in emulsions: particle size distribution and interfacial activity. *Food Hydrocolloids*, 533-542.

Hubbermann, E. M., Heins, A., Stockmann, H. and Schwarz, K. (2006). Influence of acids, salt, sugars and hydrocolloids on the colour stability of anthocyanin rich black currant and elderberry concentrates. *European Food Research and Technology*, 83-90.

Husni, A., Subaryono, S., Pranoto, Y., Taswir, T. and Ustadi, U. (2012). Pengembangan metode ekstraksi alginat dari rumput laut *Sargassum* sp. sebagai bahan pengental (Development of Alginate Extraction Method from *Sargassum* sp. as Thickening). *Agritech*, 1.

Imamura, Y., Mukohara, T., Shimono, Y., Funakoshi, Y., Chayahara, N., Toyoda, M., Kiyota, N., Takao, S., Kono, S., Nakatsura, T. and Minami, H. (2015). Comparison of 2D-and 3D-culture models as drug-testing platforms in breast cancer. *Oncology Reports*, 1837-1843.

Jothisaraswathi, S., Babu, B. and Rengasamy, R. (2006). Seasonal studies on alginate and its composition II: Turbinaria conoides (J. Ag.) Kütz. (Fucales, Phaeophyceae). *Journal of Applied Phycology*, 161-166.

Jung, J. T., Han, S. Y. and Chung, E. H. (2018). Characteristic of treatment from paper mill wastewater by UASB reactor using anaerobic biobead. *KSWST Journal of Water Treatment*, 11-17.

Kailasapathy, K. (2006). Survival of free and encapsulated probiotic bacteria and their effect on the sensory properties of yoghurt. *LWT Food Science and Technology*, 1221-1227.

Kanagaraj, J. and Mandal, A. B. (2012). Combined biodegradation and ozonation for removal of tannins and dyes. *Newsletter of the Council of Scientific and Industrial Research*, 148-170.

Kang, S. M, Lee, J. H., Yun, S. H. and Takayama, S. (2021). Alginate microencapsulation for three-dimensional *in vitro* cell culture. *ACS Biomaterials Science and Engineering*, 2864-2879.

Kaur, I., Agnihotri, S. and Goyal, D. (2021). Fabrication of chitosan-alginate nanospheres for controlled release of cartap hydrochloride. *Nanotechnology*, 025701.

Kelly, B. J. and Brown, M. T. (2000). Variations in the alginate content and composition of *Durvillaea antarctica* and *D. willana* from southern New Zealand. *Journal of Applied Phycology*, 317-324.

Khan, S., Tøndervik, A., Sletta, H., Klinkenberg, G., Emanuel, C., Onsøyen, E., Myrvold, R., Howe, R. A., Walsh, T. R., Hill, K. E. and Thomas, D. W. (2012). Overcoming drug resistance with alginate oligo saccharides able to potentiate the action of selected antibiotics. *Antimicrobial Agents and Chemotherapy*, 5134-5135.

Kim, H. S., Song, M., Lee, E. J. and Shin, U. S. (2015). Injectable hydrogels derived from phosphorylated alginic acid calcium complexes. *Materials Science & Engineering C-Materials for Biological Applications*, 139-147.

Kim, Y., Yoon, K. and Ko, S. (2011). Preparation and properties of alginate superabsorbent filament fibers crosslinked with glutaraldehyde. *Journal of Applied Polymer Science*, 1797-1804.

King, A. H. (1983). Brown seaweed extracts (aldinates). In *Food Hydrocolloids*, Volum 2 (M. Glicksman, ed.), CRC pRESS Inc., Boca Raton, FL.

Knudsen, N. R., Ale, M. T., Fatemeh, A. and Meyer, A. S. (2017). Characterization of alginates from Ghanaian brown seaweeds: *Sargassum* spp. and *Padina* spp. *Food Hydrocoll*, 236-244.

Kong, H. J., Lee, K. Y. and Mooney, D. J. (2002). Decoupling the dependence of rheological/mechanical properties of hydrogels from solids concentration. *Polymer*, 6239-6246.

Kulseng, B., Skjak-Bræk, G., Ryan, L., Anderson, A., King, A., Faxvaag, A. and Terje, E. (1999). Antibodies against alginates and encapsulated porcine islet-like cell clusters. *Transplantation*, 978-984.

Kumar, S., Chauhan, N., Gopal, M., Kumar, R. and Dilbaghi, N. (2015). Development and evaluation of alginate-chitosan nanocapsules for controlled release of acetamiprid. *International Journal of Biological Macromolecules*, 631-637.

Kupper, F. C., Kloareg, B., Guern, J. and Potin, P. (2001). Oligoguluronates elicit an oxidative burst in the brown algal kelp *Laminaria digitata*. *Plant Physiology*, 278-291.

Lee, K. Y. and Mooney, D. J. (2012). Alginate: Properties and biomedical applications. *Progress in Polymer Science*, 106-126.

Leroux, M. A., Guilak, F. and Setton, L. A. (1999). Compressive and shear properties of alginate gel: Effects of sodium ions and alginate concentration. *Journal of Biomedical Materials Research*, 46-53.

Liu, S. L., Kang, M. M., Hussain, I., Li, K. W., Yao, F. and Fu, G. (2016). High mechanical strength and stability of alginate hydrogel induced by neodymium ions coordination. *Polymer Degradation and Stability*, 1-7.

Liu, W., Ju, X., Xie, R., Wang, W., Liu, Z. and Chu, L. (2020). Recent progress in preparation of functional capsule membranes based on co-extrusion minifluidic technique. *CIESC Journal*, 4365-437.

Ma, L., Cheng, C., Nie, C., He, C., Deng, J., Wang, L., Xia, Y. and Zhao, C. (2016). Anticoagulant sodium alginate sulfates and their mussel-inspired heparin-mimetic coatings. *Journal of Materials Chemistry B*, 3203-3215.

Mahmoud, G. A. and Mohamed, S. F. (2012). Removal of lead ions from aqueous solution using (sodium alginate/itaconic acid) hydrogel prepared by gamma radiation. *Australian Journal of Basic and Applied Sciences*, 262-273.

Mancini, F. and McHugh, T. H. (2000). Fruit-alginate interactions in novel restructured products. *Nahrung*, 152-157.

Mancini, F., Montanari, L., Peressini, D. and Fantozzi, P. (2002). Influence of alginate concentration and molecular weight on functional properties of mayonnaise. *LWT Food Science and Technology*, 517-525.

Mazzitelli, S., Capretto, L., Quinci, F., Piva, R. and Nastruzzi, C. (2013). Preparation of cell-encapsulation devices in confined microenvironment. *Advanced Drug Delivery Reviews*, 1533-1555.

Meng, Y., Wu, C., Zhang, J., Cao, Q., Liu, Q. and Yu, Y. (2015). Amphiphilic alginate as a drug release vehicle for water-insoluble drugs. *Colloid Journal*, 754-760.

Moe, S. T., Skjaak-Braek, G., Elgsaeter, A. and Smidsroed, O. (1993). Swelling of covalently crosslinked alginate gels: Influence of ionic solutes and nonpolar solvents. *Macromolecules*, 3589-3597.

Munarin, F., Kabelac, C. and Coulombe, K. L. K. (2021). Heparin-modified alginate microspheres enhance neovessel formation in hiPSC-derived endothelial cells and heterocellular *in vitro* models by controlled release of vascular endothelial growth factor. *Journal of Biomedical Materials Research*, 1726-1736.

Nagarajan, A., Shanmugam, A. and Zackaria, A. (2016). Mini review on alginate: Scope and future prospectives. *Journal of Algal Biomass Utilization*, 45-55.

Navratil, M., Gemeiner, P., Klein, J., Sturdik, E., Malovikova, A., Nahalka, J., Vikartovska, A., Domeny, Z. and Smogrovicova, D. (2002). Properties of hydrogel materials used for entrapment of microbial cells in production of fermented beverages. *Artificial Cells, Blood Substitutes, and Immobilization Biotechnology*, 199-218.

Norton, S. and Vuillemard, J. C. (1994). Food bioconversions and metabolite production using immobilized cell technology. *Critical Reviews in Biotechnology*, 193-224.

Oms-Oliu, G., Soliva-Fortuny, R. and Martin-Belloso, O. (2008). Using polysaccharide-based edible coatings to enhance quality and antioxidant properties of fresh-cut melon. *LWT Food Science and Technology*, 1862-1870.

Ooi, H. W., Mota, C., ten Cate, A. T., Calore, A., Moroni, L. and Baker, M. B. (2018). Thiolene alginate hydrogels as versatile bioinks for bioprinting. *Biomacromolecules*, 3390-3400.

Orive, G., Carcaboso, A. M., Hernandez, R. M., Gascon, A. R. and Pedraz, J. L. (2005). Biocompatibility evaluation of different alginates and alginate-based microcapsules. *Biomacromolecules*, 927-931.

Orive, G., Ponce, S., Hernandez, R. M., Gascon, A. R., Igartua, M. and Pedraz, J. L. (2002). Biocompatibility of microcapsules for cell immobilization elaborated with different type of alginates. *Biomaterials*, 3825-3831.

Osman, S. F., Fett, W. F. and Fishman, M. L. (1986). Exopolysaccharides of the phytopathogen *Pseudomonas syringae* pv. glycinea. *Journal of Bacteriology*, 66-71.

Oussalah, M., Caillet, S., Salmieri, S., Saucier, L. and Lacroix, M. (2007). Antimicrobial effects of alginate-based films containing essential oils on *Listeria monocytogenes* and *Salmonella typhimurium* present in bologna and ham. *Journal of Food Protection*, 901-908.

Papageorgiou, S. K., Kouvelos, E. P., Favvas, E. P., Sapalidis, A. A., Romanos, G. E. and Katsaros, F. K. (2010). Metal-carboxylate interactions in metal-alginate complexes studied with FTIR spectroscopy. *Carbohydrate Research*, 469-473.

Paraskevopoulou, A., Boskou, D. and Kiosseoglou, V. (2005). Stabilization of olive oil-lemon juice emulsion with polysaccharides. *Food Chemistry*, 627-634.

Paraskevopoulou, D., Boskou, D. and Paraskevopoulou, A. (2006). Oxidative stability of olive oil-lemon juice salad dressings stabilized with polysaccharides. *Food Chemistry*, 1197-1204.

Parreidt, S. T., Müller, K. and Schmid, M. (2018). Alginate-based edible films and coatings for food packaging applications. *Foods*, 7100170 (1-38).

Pawar, S. N. and Edgar, K. J. (2011). Chemical modification of alginates in organic solvent systems. *Biomacromolecules*, 4095-4103.

Pawar, S. N. and Edgar, K. J. (2012). Alginate derivatization: A review of chemistry, properties and applications. *Biomaterials*, 3279-3305.

Pawar, S. N. and Edgar, K. J. (2013). Alginate esters via chemoselective carboxyl group modification. *Carbohydrate Polymers*, 1288-1296.

Pereira, L. and Cotas, J. (2020). Introductory chapter: Alginates-A general overview. In *Alginates-Recent uses of this natural polymer*, IntechOpen: London, UK.

Praphakar, R. A., Munusamy, M. A., Alarfaj, A. A., Kumar, S. S. and Rajan, M. (2017). Zn^{2+} cross-linked sodium alginate-g-allylamine-mannose polymeric carrier of rifampicin for macrophage targeting tuberculosis nanotherapy. *New Journal of Chemistry*, 11324-11334.

Prodanovic, O., Spasojevic, D., Prokopijevic, M., Radotic, K., Markovic, N., Blazic, M. and Prodanovic, R. (2015). Tyramine modified alginates via periodate oxidation for peroxidase induced hydrogel formation and immobilization. *Reactive and Functional Polymers*, 77-83.

Putri, A. P., Picchioni, F., Harjanto, S. and Chalid, M. (2021). Alginate modification and lectin-conjugation approach to synthesize the mucoadhesive matrix. *Applied Science*, 11818.

Qin, Y. (2004). Gel swelling properties of alginate fibers. *Journal of Applied Polymer Science*, 1641-1645.

Rad, F. H., Ghaffari, T. and Safavi, S. H. (2010). *In vitro* evaluation of dimensional stability of alginate impressions after disinfection by spray and immersion methods. *Journal of Dental Research, Dental Clinics, Dental Prospects*, 130-135.

Raja, M., Liu, C. and Huang, Z. (2015). Nanoparticles based on oleate alginate ester as curcumin delivery aystem. *Current Drug Delivery*, 613-627.

Rashedy, S. H., Hafez, M. S. M. A. E., Dar, M. A., Cotas, J. and Pereira, L. (2021). Evaluation and characterization of alginate extracted from brown seaweed collected in the Red Sea. *Applied Sciences*, 6290.

Raus, R. A., Wan, M. and Nasaruddin, R. R. (2021). Alginate and alginate composites for biomedical applications. *Asian Journal of Pharmaceutical Sciences*, 280-306.

Regand, A. and Goff, H. D. (2003). Structure and ice recrystallization in frozen stabilized ice cream model systems. *Food Hydrocolloids*, 95-102.

Rioux, L. E., Turgeon, S. L. and Beaulieu, M. (2007). Rheological characterisation of polysaccharides extracted from brown seaweeds. *Journal of the Science of Food and Agriculture*, 1630-1638.

Rohof, W. O., Bennink, R. J., Smout, A. J., Thomas, E. and Boeckxstaens, G. E. (2013). An alginate-antacid formulation localizes to the acid pocket to reduce acid reflux in patients with gastroesophageal reflux disease. *Clinical Gastroenterology and Hepatology*, 1585-1591.

Sabra, W. and Zeng, A. P. (2009). Microbial production of alginates: Physiology and process aspects, in: Alginates: Biology and Applications. *Springer*, 153-173.

Sakiyama-Elbert, S. E. and Hubbell, J. A. (2001). **Functional biomaterials: Design of novel biomaterials**. *Annual Review of Materials Research*, 183-201.

Sanchez-Ballester, N. M., Bataille, B., Benabbas, R, Alonso, B. and Soulairol, I. (2020). Development of alginate esters as novel multifunctional excipients for direct compression. *Carbohydrate Polymers*, 116280.

Schulz, A., Gepp, M. M., Stracke, F., von Briesen, H., Neubauer, J. C., Zimmermann, H. (2019). Tyramine-conjugated alginate hydrogels as a platform for bioactive scaffolds. *Journal of Biomedical Materials Research*, 114-121.

Smidsrod, O. and Draget, K. I. (1996). Chemistry and physical properties of alginates. *Carbohydrate European*, 6-13.

Smidsrod, O. and Haug, A. (1968). A light scattering study of alginate. *Acta Chemica Scandinavica*, 797-810.

Smith, A. M. and Senior, J. J. (2021). Alginate hydrogels with tuneable properties. *Advances in Biochemical Engineering-Biotechnology*, 37-61.

Soon-Shiong, P., Otterlei, M., Skjak-Bræk, G., Smidsrød, O., Heintz, R., Lanza, P. and Terje E. (1991). An immunology basis for the fibrotic reaction to implanted microcapsules. *Transplantation Proceedings*, 758-759.

Sperger, D. M., Fu, S., Block, L. H. and Munson, E. J. (2011). Analysis of composition, molecular weight, and water content variations in sodium alginate using solid-state NMR spectroscopy. *Journal of Pharmaceutical Sciences*, 3441-3452.

Stokke, T. B., Kurt, D., Smidsrod, O., Yuguchi, Y., Urakawa, H. and Kajiwara, K. (2000). Small-angle x-ray scattering and rheological characterization of alginate gels Ca-alginate gels. *Macromolecules*, 1853-1863.

Strudart, A. R., Gonzenbach, U. T., Tervoor, E. and Gauckler, L. J. (2006). Processing routes to macroporous ceramics: A review. *Journal of the American Ceramic Society*, 1771-1789.

Suzuki, S., Kurachi, S., Wada, N. and Takahashi, K. (2021). Selective modification of aliphatic hydroxy groups in lignin using ionic liquid. *Catalysts*, 120.

Sweeney, I. R., Miraftab, M. and Collyer, G. (2012). A critical review of modern and emerging absorbent dressings used to treat exuding wounds. *International Wound Journal*, 601-612.

Tay, S. L. and Perera, C. O. (2004). Effect of 1-methylcyclopropene treatment and edible coatings on the quality of minimally processed lettuce. *Journal of Food Science*, 131-135.

Thomas, S. (1992). Observations on the fluid handling properties of alginate dressings. *Pharmacognosy Journal*, 85-851.

Tønnesen, H. H. and Karlsen, J. (2002). Alginate in drug delivery systems. *Drug Development and Industrial Pharmacy*, 621-663.

Varghese, S. and Elisseeff, J. H. (2006). Hydrogels for musculoskeletal tissue engineering. *Advances in Polymer Science*, 95-144.

Veerubhotla, K., Lee, Y. and Chi, H. L. (2021). Parametric optimization of 3D printed hydrogel-based cardiovascular stent. *Pharmaceutical Research*, 885-900.

Vijayaraghavan, G. and Shanthakumar, S. (2018). Effective removal of reactive magenta dye in textile effluent by coagulation using algal alginate. *Desalination and Water Treatment*, 22-27.

Vold, I. M. N., Vårum, K. M. E., Guibal, E. and Smidsrød, O. (2003). Binding of ions to chitosan-selectivity studies. *Carbohydrate Polymers*, 471-477.

Vos, P. D., Haan, B. D. and Schifgaarde R. V. (1997). Effect of the alginate composition on the biocompatibility of alginate-polylysine micro-capsules. *Biomaterials*, 273-278.

Vos, P. D., Lazarjani, H. A., Poncelet, D. and Faasa, M. M. (2014). Polymers in cell encapsulation from an enveloped cell perspective. *Advanced Drug Delivery Reviews*, 15-34.

Wang, H., Song, Z., Ciofu, O., Onsøyen, E., Rye, P. D. and Hoiby, N. (2016). OligoG CF-5/20 disruption of mucoid *Pseudomonas aeruginosa* biofilm in a murine lung infection model. *Antimicrobial Agents and Chemotherapy*, 2620-2626.

Wang, L., Liu, L., Holmes, J., Kerry, J. F. and Kerry, J. P. (2007). Assessment of film-forming potential and properties of protein and polysaccharide-based biopolymer films. *International Journal of Food Science and Technology*, 1128-1138.

Wedlock, D. J., Fasihuddin, B. A. and Phillips, G. O. (1986). Comparison of molecular weight determination of sodium alginate by sedimentation-diffusion and light scattering. *International Journal of Biological Macromolecules*, 57-61.

Westermeier, R., Murúa, P., Patiño, D. J., Muñoz, L., Ruiz, A. and Müller, D. G. (2012). Variations of chemical composition and energy content in natural and genetically defined cultivars of *Macrocystis* from Chile. *Journal of Applied Phycology*, 1191-1201.

Williams, D. F. (2009). On the nature of biomaterials. *Biomaterials*, 5897-59022.

Williams, J. A., Lai, C. S., Corwin, H., Ma, Y., Maki, K. C., Garleb, K. A., Wolf, B. W. (2004). Inclusion of guar gum and alginate into a crispy bar improves postprandial glycemia in humans. *The Journal of Nutrition*, 886-889.

Wu, J., Wu, Z., Zhang, R., Yuan, S., Lu, Q. and Yu, Y. (2017). Synthesis and micelle properties of the hydrophobic modified alginate. *International Journal of Polymeric Materials and Polymeric Biomaterials*, 742-747.

Yan, F., Lian, Y., Yang, G., Wang, P., Wu, C. and Chen, N. (2013). Screening of alginate lyase-producing strains and optimization of fermentation conditions. *Progress in Modern Biomedicine*, 5606-5609.

Yang, J., Jiang, B., He, W. and Xia, Y. (2012). Hydrophobically modified alginate for emulsion of oil in water. *Carbohydrate Polymers*, 1503-1506.

Yang, J., Xie, Y. and He, W. (2011). Research progress on chemical modification of alginate: A review. *Carbohydrate Polymers*, 33-39.

Yu, Y., Leng, C., Liu, Z., Jia, F., Zheng, Y., Yuan, K. and Yan, S. (2014). Reparation and characterization of biosurfactant based on hydrophobically modified alginate. *Colloid Journal*, 622-627.

Zeng, D., Hu, D. and Cheng, J. (2011). Preparation and study of a composite flocculant for papermaking wastewater treatment. *Journal of Environmental Protection*, 1370-1374.

Zhang, L., Li, X., Zhang, X., Li, Y. and Wang, L. (2021). Bacterial alginate metabolism: An important pathway for bioconversion of brown algae. *Biotechnology for Biofuels*, 158.

Biographical Sketch

Wei Liu, PhD

Affiliation: Weihai Marine Organism & Medical Technology Research Institute, Harbin Institute of Technology, Weihai 264209, P. R. China.

Education: PhD, Institute of Genetics and Developmental Biology, Chinese Academy of Sciences.

Business Address: 2 Wenhuaxi Road, Weihai 264209, People's Republic of China.

Research and Professional Experience: An assistant researcher of Yantai Coastal Zone Research Institute, Chinese Academy of Sciences in 2014, who was promoted to associate researcher in 2021 and is now an associate professor of Harbin Institute of Technology.

Professional Appointments: Associate Professor.

Honors: First prize winner of China Association for the promotion of industry university research cooperation in 2020. The award-winning project is the integration of key technologies of new marine microbial transformation and the application of large-scale health industry.

Publications from the Last 3 Years:
1. FixJ family regulator AcfR of *Azorhizobium caulinodans* involved in the symbiosis with the host plant. *BMC Microbiology*, 2021, 21, 80.
2. Biodiversity and geographic distribution of rhizobia nodulating with *Vigna minima*. *Frontiers in Microbiology*, 2021, 12, 665839.
3. CheY1 and CheY2 of *Azorhizobium caulinodans* ORS571 regulate chemotaxis and competitive colonization with the host plant. *Applied and Environmental Microbiology*, 2020, 86(15), e00599-20.
4. LuxR-type regulator AclR1 of *Azorhizobium caulinodans* regulates cyclic di-GMP and numerous phenotypes in free-living and symbiotic states. *Molecular Plant-Microbe Interactions*, 2020, 33(3), 528-538.
5. Diverse genomic backgrounds Vs. highly conserved symbiotic genes in *Sesbania*-nodulating bacteria: Shaping of the rhizobial community by host and soil properties. *Microbial Ecology*, 2020, 80, 158-168.
6. Long-term monoculture reduces the symbiotic rhizobial biodiversity of peanut. *Systematic and Applied Microbiology*, 2020, 43(5), 126101.
7. The genome of *Ensifer alkalisoli* YIC4027 provides insights for host specificity and environmental adaptations. *BMC Genomics*, 2019, 20, 643.
8. Prediction and functional analysis of GGDEF/EAL domain-containing proteins in *Azorhizobium caulinodans* ORS571. *Acta Microbiologica Sinica*, 2019, 59(10), 2000-2012.
9. Quorum sensing of bacteria and its cross-border signal regulation to plants. *Coast Science*, 2019, 6(1), 44-50.
10. Diversity and evolution of *Aeschynomene indica* nodulating rhizobia without the Nod factor synthesis genes in Shandong Peninsula, China. *Applied and Environmental Microbiology*, 2019, 3085(22), e00782-19.

Hui-Jing Li, PhD

Affiliation: Weihai Marine Organism & Medical Technology Research Institute, Harbin Institute of Technology, Weihai 264209, P. R. China.

Education: PhD, Institute of Chemistry, Chinese Academy of Sciences.

Business Address: 2 Wenhuaxi Road, Weihai 264209, People's Republic of China

Research and Professional Experience: A postdoctoral at Oregon State University and Université Paris from 2004 to 2011, who served as an associate professor of Harbin Institute of Technology in 2011 and was promoted to Professor in 2014.

Professional Appointments: Professor.

Honors: First prize winner of China Chamber of Commerce in 2021. The award-winning project is the research and application of natural carbohydrate functional factors and key technologies for efficient biomimetic preparation.

Publications from the Last 3 Years:
1. Immunological effect of fucosylated chondroitin sulfate and its oligomers from Holothuria fuscogilva on RAW 264.7 cells, *Carbohydrate Polymers*, 2022, 287, 119362.
2. Foxtail millet prolamin as an effective encapsulant deliver curcumin by fabricating caseinate stabilized composite nanoparticles, *Food Chemistry*, 2022, 367, 130764.
3. Co-assembly of foxtail millet prolamin-lecithin/alginate sodium in citric acid-potassium phosphate buffer for delivery of quercetin, *Food Chemistry*, 2022, 381, 132268.
4. Chondroitin sulfate deposited on the foxtail millet prolamin/caseinate nanoparticles to improve physicochemical properties and enhance therapeutic effect, *Food & Function*, 2022, 13 (9), 5343–5352.
5. A specifically triggered turn-on fluorescent sensor platform and its visual imaging of HClO in cells, arthritis and tumors, *Journal of Hazardous Materials*, 2022, 427, 127874.

6. A water-soluble turn-on fluorescent probe for rapid discrimination and imaging of Cys/Hcy and GSH in cells and zebrafish through different fluorescent channels, *Dyes and Pigments*, 2022, 199, 110058.
7. Fabricating of grape seed proanthocyanidins loaded zein-NaCas composite nanoparticles to exert effective inhibition of Q235 steel corrosion in seawater, *Journal of Molecular Liquids*, 2022, 348, 118467.
8. Structural elucidation and antidiabetic activity of fucosylated chondroitin sulfate from sea cucumber *Stichopus japonicus*, *Carbohydrate Polymers*, 2021, 262, 117969.
9. Inhibition of mild steel corrosion in 1 M HCl by chondroitin sulfate and its synergistic effect with sodium alginate, *Carbohydrate Polymers*, 2021, 260, 117842.
10. Simple turn-on fluorescent sensor for discriminating Cys/Hcy and GSH from different fluorescent signals, *Analytical Chemistry*, 2021, 93 (4), 2244-2253.

Yan-Chao Wu, PhD

Affiliation: Weihai Marine Organism & Medical Technology Research Institute, Harbin Institute of Technology, Weihai 264209, P. R. China.

Education: PhD, Institute of Chemistry, Chinese Academy of Sciences.

Business Address: 2 Wenhuaxi Road, Weihai 264209, People's Republic of China.

Research and Professional Experience: A postdoctoral at Institut de Chimie des Substances Naturelles, CNRS from 2006 to 2011, who served as a professor of Harbin Institute of Technology in 2011.

Professional Appointments: Professor.

Honors: Taishan industrial leading talents in Shandong Province, China.

Publications from the Last 3 Years:
1. Mesoporous crosslinked chitosan-activated clinoptilolite biocomposite for the removal of anionic and cationic dyes, *Colloids and Surfaces B-Biointerfaces*, 2022, 216, 112579.

2. Corrosion resistance and antibacterial activity of procyanidin B2 as a novel environment-friendly inhibitor for Q235 steel in 1 M HCl solution, *Bioelectrochemistry*, 2022, 143, 107969.
3. Highly effective Q235 steel corrosion inhibition in 1M HCl solution by novel green strictosamide from *Uncaria laevigata*: Experimental and theoretical approaches, *Journal of Environmental Chemical Engineering*, 2022, 10 (3), 107581.
4. Switchable and efficient conversion of 2-amido-aryl oxazolines to quinazolin-4(3*H*)-ones and N-(2-chloroethyl) benzamides, *Organic Chemistry Frontiers*, 2021, 8 (3), 584-590.
5. Synthesis of difluorinated 3-oxo-N,3-diarylpropanamides from 4-arylamino coumarins mediated by selectfluor, *Organic Chemistry Frontiers*, 2021, 8 (23), 6636-6641.
6. Structure-based discovery of novel 4-(2-fluorophenoxy)quinoline derivatives as c-Met inhibitors using isocyanide-involved multicomponent reactions, *European Journal of Medicinal Chemistry*, 2020, *193*, 112241.
7. Design, synthesis and biological evaluation of novel N-sulfonylamidine-based derivatives as c-Met inhibitors via Cu-catalyzed three-component reaction, *European Journal of Medicinal Chemistry*, 2020, *200*, 112470.
8. Release of antidiabetic peptides from Stichopus japonicas by simulated gastrointestinal digestion, *Food Chemistry*, 2020, 315, 126273.
9. Performance and mechanism of a composite scaling–corrosion inhibitor used in seawater: 10-methylacridinium iodide and sodium citrate, *Desalination*, 2020, 486, 114482.
10. Fructan from *Polygonatum cyrtonema* Hua as an eco-friendly corrosion inhibitor for mild steel in HCl media, *Carbohydrate Polymers*, 2020, 238, 116216.

Chapter 2

Submicron and Nano-Sized Gel Particles Based on Alginate and Sulfated Alginate for Protein Encapsulation

A. Anitha, PhD, Jiankun Yang, Yingjun Gao, PhD, Auriane Gueguen, Broden Diggle and Lisbeth Grøndahl[*], PhD

School of Chemistry and Molecular Biosciences, The University of Queensland, Brisbane, Australia

Abstract

Alginate (ALG) and its heparin (HEP) mimetic derivative sulfated alginate (S-ALG) have been widely studied for use in various biomedical applications. This chapter gives an overview into fabrication of ALG- and S-ALG-based submicron- and nano-sized gel particles with a focus on protein encapsulation. Include is our research on the optimisation of gel particles composed of ALG or S-ALG for the encapsulation of the high pI protein Lactoferrin (Lf). The binding of Lf to ALG and S-ALG as studied through surface plasmon resonance (SPR) revealed non-specific binding to ALG while for S-ALG an association constant, $K_A = 2.0 \times 10^7$ M^{-1}, was determined. The gel particles were synthesised using a nanoprecipitation method and crosslinked with chitosan (CHI) and calcium ions. The particle size characterised using dynamic light scattering (DLS) and nanoparticle tracking analysis (NTA) was in the range of 200–600 nm depending on the reaction conditions. Encapsulation of Lf reduced the particle size for all particles and calcium ions for ionic cross linking was only required for Lf encapsulation in

---

[*] Corresponding Author's Email: l.grondahl@uq.edu.au.

In: Properties and Applications of Alginate
Editor: Michael Y. Wilkerson
ISBN: 979-8-88697-371-6
© 2022 Nova Science Publishers, Inc.

ALG-based gel particles. Suspension stability was evaluated through DLS measurements revealing that only particles produced using nanoprecipitation and sonication had low polydispersity over a 6-day period.

Keywords: surface plasmon resonance, lactoferrin, alginate sulfate, particle size distribution and suspension stability

Introduction

Alginate (ALG) is an anionic polysaccharide commonly extracted from brown algae. This linear copolymer is composed of β-D-mannuronate (M) and α-L-guluronate (G) residues linked by (1,4)-glycosidic linkages (Grøndahl et al. 2020). Due to its biocompatibility with many tissues, low cost, and non-toxicity, ALG is a useful material for a variety of biomedical applications, including drug encapsulation (Lee and Mooney 2012; Pawar and Edgar 2012; Grøndahl et al. 2020). ALG has the ability to form a gel under mild conditions such as in the presence of calcium ions facilitated by electrostatic interactions between the divalent cation and the anionic polysaccharide (Lee and Mooney 2012; Pawar and Edgar 2012; Grøndahl et al. 2020). Likewise, ALG can interact electrostatically with positively charged polymers such as chitosan (CHI) to form a polyelectrolyte complex (Lawrie et al. 2007; Aston et al. 2015; Bourganis et al. 2017).

The C2 and/or C3 sulfated analogue of alginate (S-ALG) has been shown to be an effective heparin (HEP) mimetic which binds strongly to certain HEP-binding proteins (Arlov and Skjåk-Bræk 2017). This is attributed to the additional negative charges of S-ALG, which facilitate stronger electrostatic interactions with a range of cationic surfaces, including proteins (Pawar and Edgar 2012; Arlov and Skjåk-Bræk 2017). These charged sulfate groups also result in tighter cross-linking interactions and higher solubility in water than ALG (Pawar and Edgar 2012; Arlov and Skjåk-Bræk 2017).

In this chapter the focus is on ALG and S-ALG based gel particle systems in the nano- and sub-micron-size range which have been used for drug delivery. According to the National Science and Technology Council (NSTC) Committee on Technology, subcommittee on Nanoscale Science Engineering and Technology (NSET), nanotechnology is defined as the development of research and technology at a critical length scale of 1–100 nm. However, in particular circumstances it is acceptable to use sizes up to 300 nm, particularly when referring to polymeric nanoparticles (Hornyak et al. 2008). Therefore, in this chapter, the term nanosized particles is considered for particles up to 300 nm in diameter and particles in the range of 301–999 nm are considered as sub-micron sized particles.

Various methods have been used for the fabrication of ALG- and S-ALG-based submicron and nano-sized particles for drug delivery with examples listed in Table 1. All examples listed involve ionic rather than covalent crosslinking as this represents the most common approach. Three different methods have been explored in multiple studies. These are nanoprecipitation, emulsion (w/o or w/o/w), and sonication. In addition, the combination of sonication and nanoprecipitation has been reported. In addition to the use of ALG, many formulations also include a positively charged polyelectrolyte such as CHI, trimethyl-CHI (TMC), succinyl-CHI (SCS), or poly-L-lysine (PLL) allowing for polyelectrolyte complex formation with the negatively charged ALG polymer. For the emulsion formulations, stabilisers such as Tween 80, mineral oil, polyvinyl alcohol (PVA) or caster oil are used. In addition, one study produced core-shell particles with polycaprolactone (PCL) as the shell. A range of particle sizes have been reported, the smallest achieved using a w/o emulsion system (65 nm) or sonication (90 nm).

A range of drug types have been encapsulated in these particles. Some are small molecule drugs such as Doxorubicin, some are peptides such as Liraglutide, while a large proportion of the studies are concerned with encapsulation of protein drugs. These include connective tissue growth factor (CTGF) and insulin-like growth factor (IGF-I), and insulin. Furthermore, S-ALG has been explored for encapsulation of a series of HEP-binding proteins.

The current study evaluates ALG- and S-ALG-based particles for encapsulation of Lactoferrin (Lf), a non-heme, iron-binding, high pI glycoprotein which belongs to the transferrin family (Legrand et al. 2008). Lf is composed of a single polypeptide chain folded into two homologous lobes. A fully iron saturated Lf protein binds two ferric ions (Fe(III)) (Steijns and Van Hooijdonk 2000). The properties of Lf depends on the level of ion saturation with the fully saturated protein displaying the highest thermal stability (Bokkhim et al. 2013; Bokkhim et al. 2014). Lf is known for multiple health benefits including anti-carcinogenic, immunomodulatory, anti-inflammatory and antimicrobial effects (Legrand et al. 2008). Many studies have therefore investigated encapsulation of Lf for applications in cancer therapy, osteoarthritis therapy, Parkinson treatment, as a dental antibacterial agent, and for tendinitis anti-inflammation (Martınez-Gomis et al. 1999; Samarasinghe, Kanwar, and Kanwar 2014; Tammam, Azzazy, and Lamprecht 2018; H.J. Choi et al. 2020; Xiong et al. 2020).

Table 1. Examples of studies reporting fabrication of drug-loaded nano and submicron sized gel particles from ALG and/or S-ALG

Fabrication method	Polymer(s)	Encapsulated Drug	Particle size (method of determination)	Ref
Nanoprecipitation	ALG	Heat-labile B subunit	144 nm (DLS)	(Kordbacheh et al. 2018)
Nanoprecipitation	ALG, CHI	Nisin	110 nm (AFM)	(Zimet et al. 2018)
Nanoprecipitation	ALG[a], PLL	TGF-β3	377 nm (DLS)	(Roger et al. 2020)
Nanoprecipitation	ALG[b], CHI	Amygdalin	119 nm (DLS) 41 nm (SEM)	(Sohail and Abbas 2020)
Nanoprecipitation	ALG, CHI, albumin	Insulin	139–534 nm (DLS)	(Hadiya et al. 2021)
Nanoprecipitation	S-ALG	bFGF[j], VEGF[k], PDGF-ββ[l], HGF[m]	100–440 nm (NTA)	(Ruvinov et al. 2016)
Nanoprecipitation	ALG, S-ALG[c]	Salmon calcitonin	323–771 nm (DLS)	(Chen et al. 2021)
Nanoprecipitation	ALG[d], CHI	Insulin	423–850 nm (DLS)	(Sarmento et al. 2006)
Nanoprecipitation	ALG, TMC	Insulin	370 nm (DLS)	(Mansourpour et al. 2015)
w/o emulsion	ALG[e]	Urease	65 nm (DLS)	(Saxena et al. 2017)
w/o/w emulsion	ALG	Plasmid DNA	463 nm (DLS)	(Y.H. Choi et al. 2017)
w/o/w emulsion	S-ALG[f], PCL, PVA	CTGF, IGF-I	200 nm (DLS)	(Maatouk et al. 2021)
w/o emulsion	ALG[g], aloe vera	Insulin	480 nm (DLS) 400 nm (SEM)	(Basha et al. 2021)
Sonication	ALG[h], SCS	Quercetin	90 nm (DLS)	(Mukhopadhyay et al. 2018)
Sonication and nanoprecipitation	ALG[i], CHI	Doxorubicin	300 nm (DLS)	(Yoncheva et al. 2019)
Sonication and nanoprecipitation	ALG, CHI	Liraglutide	<100 nm (SEM) 600 nm (DLS)	(Shamekhi, Tamjid, and Khajeh 2021)

[a] M_w 70–80 kDa; [b] medium viscosity, M_w 75–100 kDa; [c] DS = 1.34; [d] low viscosity, F_G = 0.39; [e] M_w 120–190 kDa; [f] DS = 1, synthesised from ALG with low viscosity M_w 75–200 kDa; [g] M_w 3 kDa; [h] M:G = 6.5:3.6; [i] very low viscosity; [j] basic fibroblast growth factor; [k] vascular endothelial growth factor; [l] platelet-derived growth factor-ββ; [m] hepatocyte growth factor.

Encapsulation of Lf in alginate-based matrices has previously been achieved in large ALG gel particles (Bokkhim et al. 2016a), micron-sized ALG gel particles (Bokkhim et al. 2016b), and in ALG/CHI nanocarriers

(Samarasinghe, Kanwar, and Kanwar 2014). Furthermore, a range of proteins have been encapsulated in ALG/CHI polyelectrolyte complex fibers which showed enhanced control over the release rate for both high and low p*I* proteins and this could be further modulated by inclusion of HEP (Liao et al. 2005).

The current study evaluates the effect of the degree of ALG sulfation and fabrication parameters on the size distribution and suspension stability of ALG- and S-ALG-based gel particles encapsulating Lf and strengthen using polyelectrolyte complex interactions with CHI. The binding affinity of ALG and S-ALG for Lf was evaluated using surface plasmon resonance (SPR) and correlated to the particle properties.

Method

Materials

Alginic acid sodium salt from brown algae (medium viscosity, M_w 280 kDa, M/G ratio of 1.5) (ALG) was obtained from Sigma Aldrich. Biacore™ Sensor Chips (SA) were from GE Healthcare Bio-Sciences AB. Chitosan PROTOSAN CL113 (viscosity < 20 cp, M_w < 200 kDa) was purchased from NovaMatrix. Bovine lactoferrin (NatraFerrin, iron saturation 13%) was provided by MG Nutritionals®. Deionised (DI) water was obtained through a VWR International PureLab Flex ELGA purification system (conductivity 18.2 mΩ cm) and used throughout. HBS-EP buffer (pH 8.9) was prepared from NaCl 300 mM, ethylenediaminetetraacetic acid (EDTA, 3 mM), 4-(2-hydroxyethyl)-1-piperazineethanesulfonic acid (HEPES, 10 mM), and Tween 20 (0.05%). The buffer system made was filtered through a 0.2 μm syringe filter and all the solutions used for Biacore™ was degassed for 10 minutes prior to use.

Synthesis of S-ALG

Two different methods were used to synthesise S-ALG, both using the tributyl ammonium salt of ALG (T-ALG) as the starting material to enhance the solubility of ALG in dimethylformamide (DMF). The synthesis of T-ALG followed a procedure by Freeman et al. (Freeman, Kedem, and Cohen 2008).

Briefly, sodium alginate (1.0 g) was dissolved in water (300 mL) overnight. The ALG solution was cooled in an iced bath, added Dowex ion exchange resin and stirred for 5 minutes. After removing the resin by paper filtration, the pH of the filtrate was adjusted to 6–6.5 through the addition of tributylamine (TBA) (approximately 2.5 mL) followed by stirring under vacuum for 4 hours to remove excess TBA. The product was collected through lyophilisation and characterised through elemental analysis (N% = 2.0–3.6) and FTIR spectroscopy (C-H stretch: 2875–2965 cm^{-1}).

Method-1 for the synthesis of S-ALG used H_2SO_4/N,N'-dicyclohexyl-carbodiimide (DCC) as previously described (Freeman, Kedem, and Cohen 2008). Briefly, a mixture of dry DMF (100 mL) and sulfuric acid (1–3 g) were added T-ALG (N% of 2.0–2.8; 0.3 g) and DCC (6–7 g) at room temperature, and the mixture was stirred for 2 hours. The precipitated dicyclohexylurea biproduct was removed by vacuum filtration. Dichloromethane (DCM, 300 mL) was added to the filtrate causing precipitation of the TBA salt of S-ALG which was isolated and added to a NaOH solution (0.5 M, 150 mL) and stored in a fridge overnight. Isolation of the sodium salt of S-ALG was achieved through dialysis against water followed by lyophilisation.

Method-2 for the synthesis of S-ALG using sulfur trioxide pyridine complex (SO_3.py) as the sulfating agent as previously described by Mhanna et al. (Mhanna et al. 2014). Purified and dried SO_3.py complex (0.7–2.8 g) was added to a solution of T-ALG (0.3 g) in anhydrous DMF (35 mL) and stirred for 4 hours. This solution was then poured into acetone (200 mL) with stirring to precipitate the S-ALG and added ethanolic NaOH (0.1 M) to adjust the pH to 7–8. The product was isolated by filtration, dialysis and lyophilisation.

The S-ALG products were characterised by elemental analysis (S% = 7.0–12.8; N% < 0.3) and FTIR spectroscopy (S=O stretch at 1230 cm^{-1}). The degree of substitution (DS, number of sulfate groups per uronic acid monomer) was determined for individual samples using the equation below.

$$DS = \frac{\%S}{M(S)} \times \frac{M(C) \times 6}{\%C}$$

where %S and %C are the mass % from elemental analysis, M(S) and M(C) are the molar masses of sulfur and carbon.

Samples are named according to the method of preparation and the DS value, for example, an S-ALG sample with a DS of 1.1 prepared using method-1 is named S-ALG1(1.1).

Synthesis of Biotinylated ALG and S-ALG

The biotinylation of ALG and S-ALG1(0.6) was carried out based on a published method (Polyak, Geresh, and Marks 2004) using 1-ethyl-3-(3'-dimethylaminopropyl) carbodiimide HCl (EDC)/N-hydroxysuccinimide (NHS) chemistry. ALG and S-ALG1(0.6) (100 mg) were each dissolved in 2-(N-morpholino) ethane sulfonic acid (MES) buffer (0.1 M, 10 mL, pH 6) overnight. Biotin hydrazide (25 mg) was added to the polymer solution and left stirring for one hour to facilitate homogeneous dispersion of the biotinylating reagent. Then, NHS (19.2 mg) and EDC (10.8 mg) was added and left stirring for 3 hours at room temperature. The biotinylated products were isolated by dialysis and lyophilisation and stored in a desiccator kept in the fridge until use.

Evaluation of the Interaction between the Biotinylated Polymers and Lf

The Biacore™ 3000 was used for evaluating the interaction between Lf and ALG or S-ALG1(0.6). For this an optical biosensor equipped with Biacore™Sensor Chip (SA) along with the biotinylated polymers were used. Immobilisation of biotinylated ALG (200 ng/mL, HBS-EP buffer, pH 8.9) was done at a flow rate of 30 µL/min for 0.2 minutes resulting in a response of 3.3 RU. Immobilisation of biotinylated S-ALG1(0.6) (20,000 ng/mL, HBS-EP buffer, pH 8.9) was done at a flow rate of 30 µL/min for 2 minutes resulted in a response of 45.5 RU. In both cases, the running buffer was HBS-EP (pH 8.9) with 0.05% Tween 20. After immobilisation, the SA chip was primed twice by injecting the running buffer ensuring a stable baseline.

The measurement was done using Lf solutions (25 to 300 nM, HBS-EP buffer, pH 8.9) injected onto the SA chip at 25 °C at a flow rate of 30 µL/min for 4 minutes (association phase). This concentration range was found to include solutions with concentrations more than $2 \times K_D$ as recommended by the manufacturer. Following each Lf injection, the surface was regenerated using a 3.5 M $MgCl_2$ solution for 4 minutes at a flow rate of 30 µL/min. The individual response in each of the Lf injections was obtained by the difference in response between the sample and reference flow cells. These response differences were used to obtain the affinity constant using the linear form of the Langmuir equation ($R^2 = 0.995$) based on the 1:1 Langmuir binding model.

Fabrication of Gel Particles

The overall approach to gel particle fabrication is shown in Figure 1. The variables evaluated are highlighted and include the type of ALG or S-ALG polymer, the energy input (stirring or sonication), the isolation procedure (filtration or centrifugation), and the characterisation method (DLS or NTA). Furthermore, the amount of Lf and the presence of calcium was also evaluated. The descriptions below are divided according to the energy input used; nanoprecipitation (stirring only) or nanoprecipitation and sonication.

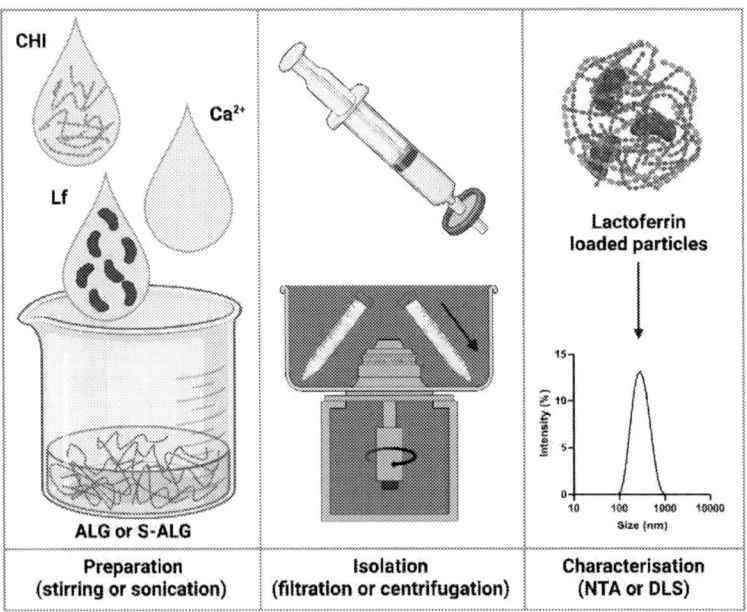

Figure 1. Overall schematic diagram of gel particle preparation, isolation and characterisation.

Nanoprecipitation

To an ALG or S-ALG solution (6 mL, 1 g L^{-1}), Lf (3 mL, 0 or 0.2 or 0.4 g L^{-1}), CaCl$_2$ (1 mL, 0 or 0.5 g L^{-1}) and CHI (1.5 mL, 0.5 g L^{-1}) were added dropwise under stirring (1250 rpm) in a sequential manner. Each solution was added over 20 minutes followed by a further 30 minutes of stirring. The volume of the ALG or S-ALG solution varied between different experiments using 3, 6 or 18 mL, however, the reagent ratio was kept constant for each type of particle. No difference in particle size was observed using different total volume.

Nanoprecipitation and Sonication

The order of addition of the solutions in the nanoprecipitation and sonication method were same as described above. Lf (3 mL, 0 or 0.2 or 0.4 g L^{-1}), CaCl$_2$ (1 mL, 0 or 0.5 g L^{-1}) and CHI (1.5 mL, 0.5 g L^{-1}) were added to an ALG or S-ALG solution (6 mL, 1 g L^{-1}) sequentially. Each solution was dropwise added over 5 minutes of 25 W probe-type sonication (pulse time: 20 seconds on and 10 seconds off). A further 5 minutes of sonication was then applied to the system.

Isolation

Isolation of the particles was achieved using filtration or centrifugation. For filtration, the suspension was filtered through a 5 μm Fluoropore™ PTFE filter (hydrophobic nonsterile PTFE syringe filter, SLLS025NS, from Millipore Millex) and the filtrate was collected. For centrifugation, the suspension was centrifuged at 2,500 g (bench top centrifuge) for 10 minutes and the supernatant collected.

Lactoferrin (Lf) Stability

The circular dichroism (CD) spectra of Lf were used to evaluate the Lf stability after exposure to heat or sonication. To prepare heat treated Lf (Lf-H), the protein solution (0.2 g L^{-1} or 0.5 g L^{-1} for 200–250 nm and 250–600 nm wavelength ranges, respectively) was heated for 20 minutes at 55 °C with constant stirring. To prepare sonicated Lf (Lf-S), the protein solution (0.1 g L^{-1} for 200–250 nm, 5 g L^{-1} for 250–350 nm and 300–600 nm wavelength ranges) was exposed to 25 W of probe-type sonication (pulse time: 20 seconds on and 10 seconds off) for 20 minutes. The spectra were obtained in aqueous solution at room temperature by a CD-spectropolarimeter (Jasco J-715 spectropolarimeter with a J-700 Spectra manager software) following previous work (Bokkhim et al. 2013). The spectra were obtained for three wavelength regions of 200–250 nm, 250–350 nm and 300–600 nm. The ellipticities are expressed in mdeg directly obtained from the instrument.

Characterisation

Dynamic Light Scattering (DLS) was preformed using a Zetasizer Ultra or Nano-ZS or 3000 HS (Malvern, UK) at 25°C. The former two had a higher resolution of 70 size bins as opposed to 24 size bins for the latter. Any difference in intensity weighted mean particle size where particles were prepared with the same conditions is likely due to the less resolved number weighted particle size distribution (PSD) achieved using the 3000HS (Garms et al. 2021). The refractive index of the dispersing medium was set to 1.33. Gel particle suspensions were ensured to have a count rate below 500 and analysed in a disposable plastic cuvette after synthesis or no later than 2 days after isolation. The size measurements were taken with 3 repeat measurements each of 10–15 cycles and are reported as z-average values unless otherwise stated.

The suspension stability of the gel particles in water was evaluated for six days at 37°C using a shaker water bath with 90 strokes per minute. The size distribution was evaluated daily using DLS.

Nanoparticle Tracking Analysis (NTA) measurements were performed with a NanoSight NS300 (Malvern, UK) at 25 °C. The particle suspensions were diluted in water and injected with a sterile syringe (1 mL) controlled by the NanoSight syringe pump. For each sample, three videos were recorded and analysed by the software. Size distribution curves were developed as a concentration (particles per mL) distribution across bin widths (5 nm).

The Zetasizer Ultra or Nano-ZS were used to measure the zeta potential at 25°C of the particles suspended in 10 mM NaCl solution loaded into DTS1070 Folded Capillary Zeta Cell (Malvern, UK) for measurement.

Statistical Analysis

The data are presented as mean ± standard deviation. To compare two groups, statistical differences were evaluated with an Anova: single factor method. The statistical significance criterion of data means was $p \leq 0.05$.

Results and Discussion

Evaluation of the Binding Affinity of ALG and S-ALG to Lf

The use of SPR to evaluate the binding affinity of proteins with polyelectrolytes is a common approach. In this manner, a comparison between HEP (DS = 1.5 normalised per monomer unit) and S-ALG (DS = 0.8 calculated from information provided) binding to a range of HEP-binding growth factors was evaluated by Freeman et al. (Freeman, Kedem, and Cohen 2008; Freeman and Cohen 2009). Their studies revealed that some HEP binding proteins (e.g., epidermal growth factor (EGF), HGF, IGF and interleukin 6 (IL-6)) displayed similar binding affinities to HEP as S-ALG while others (e.g., PDGF-ββ, stromal cell-derived factor 1 (SDF-1), and VEGF) showed higher binding constants (K_A) for S-ALG than HEP and others again, (bFGF and acidic fibroblast growth factor (aFGF)) exhibited stronger binding to HEP than S-ALG. A possible explanation for these observations is that for bFGF and aFGF, the interaction between the proteins and HEP is dominated by specific hydrogen bonding and Van der Waals interactions via multifunctional binding sites on the HEP backbones rather than from electrostatic interactions which is likely the dominant binding mode for S-ALG (Kamei et al. 2001). However, analysing the data obtained by Freeman et al. it is apparent that there is no clear correlation between the protein p*I* and binding affinity or specificity towards either HEP or S-ALG which could indicate more complex interactions between S-ALG and the proteins.

Freeman et al. furthermore reported that ALG showed only non-specific binding to these proteins (Freeman, Kedem, and Cohen 2008). This is supported by other researchers. Schmidt et al. demonstrated through SPR measurements that the association between vascular endothelial growth factor receptor (VEGFR) and VEGF is more favorable in the presence of S-ALG (DS = 0.05–0.13) compared to ALG-VEGF or VEGF control samples (Schmidt et al. 2016). Mhanna et al. showed through ELISA assays a significant difference in the binding of S-ALG with bFGF compared to ALG and that the binding affinity increased with the sulfate content of S-ALG (DS 0.8 to 2.6) (Mhanna et al. 2017). This study thus supports that the interactions between bFGF and S-ALG is dominated by electrostatic interactions.

In the current study, Lf binding with ALG or S-ALG1(0.6) immobilised on a SA chip was analysed using SPR. A typical SPR response plot for S-ALG1(0.6) is shown in Figure 2A and displays a ligand association phase reaching maximum adsorption on to the surface followed by a decline upon addition of the running buffer. In order to regenerate the SA chip between runs, a regeneration solution was used to achieve baseline response prior to the subsequent binding study. It was observed that Lf showed non-specific binding to ALG as described above. In contrast, S-ALG1(0.6) displayed specific binding to Lf (Figure 2) and when applying the 1:1 Langmuir binding model a value for the association constant of $K_A = 2.0 \times 10^7$ M^{-1} was determined. This indicates a strong binding affinity between Lf and S-ALG1(0.6) which has the potential to prolong the Lf stability upon encapsulation in an S-ALG matrix (Re'em and Cohen 2011).

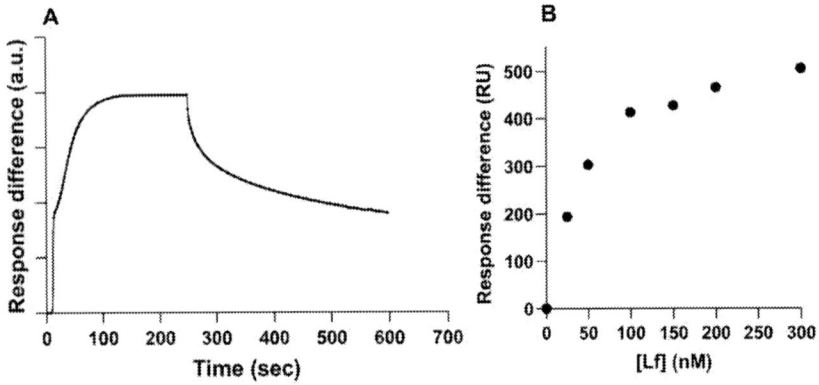

Figure 2. S-ALG1(0.6)-Lf interaction evaluated using SPR. (A) represents a typical Biacore™ response plot; (B) response difference plotted against the Lf concentration.

Reported K_A values for Lf binding with HEP or heparan sulfate (HS, DS 1.0–1.4 per disaccharide unit) are tabulated in Table 2. It has previously been shown that the mode of polysaccharide attachment onto the SA chip affects the measured binding constant (Osmond et al. 2002) and therefore Table 2 only includes examples for which it was done in the same way as in the current study. Previous works have determined values in the range $1 \times 10^5 – 2.3 \times 10^7$ M^{-1}. The relatively large range in the reported values for HEP and HS binding to Lf can be attributed to differences between the studies including the running buffer (pH and ionic strength), molecular weight of HEP, the type of Lf. The value obtained in the current study is similar to the value obtained by Nagagawa et al. for HEP – Lf binding.

Table 2. Reported K_A values for Lf binding with HEP or HS

Method (Buffer, pH)	Lf (Iron content)	Polymer (M_w)	K_A (M^{-1})	Reference
Biacore™ 3000 (HBS-EP with 0.05% Tween 20, 8.9)	Bovine (13%)	S-ALG (-)	2.0×10^7	This work
Biosensor (PBS, 7.4)	Human	HEP	1.6×10^6	(C.J. Choi et al. 2009)
Biacore™ 3000 (PB[a]/NaCl, 7.2)	Human	HEP (12 kDa)	4×10^6	(Kumar et al. 2012)
Biacore™ (HBS-P, 7.4)	Human	HS (6 kDa)	1.0×10^5	(Duchardt et al. 2009)
Biacore™ 2000 (HBS–EP, 7.4)	-	HEP (-)	2.3×10^7	(Nakagawa et al. 2009)

[a] Phosphate buffer.

Synthesis and Characterisation of Gel Particles

The method used in the current study to produce the ALG and S-ALG based gel particles was based on a method reported by Rajaonarivony et al. (Rajaonarivony et al. 1993). In their study they used ALG (M_w 75–100 kDa, M:G = 3:2) as the main matrix material and PLL as the positive polyelectrolyte to form the particles. They investigated many different variables, including the calcium concentration, the ALG concentration, and the sequence by which the different compounds were added. They determined the sol-gel transition to occurred at a calcium concentration of 3.6 mM (at an ALG concentration of 0.5% w/v). In addition, they observed that for calcium concentrations of 0.9 or 1.8 mM the system had lower viscosity than the pure ALG solution. This was attributed to the rearrangement of the ALG polymers induced by calcium ions forming micro-domains with high local concentration instead of a network of polymer, a phenomenon commonly described as a pre-gel state (Abied, Brulet, and Guenet 1990). Regarding the ALG concentration, they observed that at a low concentration (0.06% w/v) induced micro-gel formation rather than formation of a macro-gel. For the sequence of adding the different components, they found that the first compound added to ALG solution strongly impacted the particle size. When they added Doxorubicin or PLL before Ca^{2+}, the particle size was enlarged. The optimal sequence was found to be addition of Ca^{2+} prior to the drug followed by PLL. This study has been cited by 279 papers and 86 papers referred to their method of preparing gel particles (evaluated June 2022).

In the current study, we chose to add Lf prior to Ca^{2+} as they can both act as crosslinkers, however, we otherwise followed the recommendations of the previous study. We evaluated a number of parameters for the fabrication of sub-micron and nano-sized gel particles. These included the type of ALG (different degrees of sulfation), the addition of calcium, the amount of Lf and the isolation protocol (filtration or centrifugation). We also evaluated the use of nanoprecipitation and a combination of nanoprecipitation and sonication.

First Series: Nanoprecipitation

The first series of experiments used nanoprecipitation under stirring, filtration for isolation, the polymers ALG and S-ALG1(1.1), and different amounts of Lf. It also evaluated if the addition of calcium was required when the high p*I* protein Lf was encapsulated. For this study, the particle size is reported as the z-average obtained from DLS measurements. The z-average size data is presented in Figure 3A. This series of experiments were carried out by two different researchers and agreement in the resulting data was observed.

The empty ALG particles had a size of 561 nm, whereas ALG particles with 10 and 20% loading showed a significantly reduced size of 438 and 355 nm, respectively. For the S-ALG1(1.1) particles, the empty particles likewise showed a significantly larger size (334 nm) compared to the Lf loaded particles (241 and 216 nm for 10 and 20% Lf, respectively) (Figure 3A). Lf is a cationic protein at neutral pH, and it has been shown that can form electrostatic interactions with the anionic polymer ALG (Bokkhim et al. 2016a). Furthermore, strong binding affinity between S-ALG1(0.6) and Lf was demonstrated in the current study. The reduced particle size in the presence of the protein cargo can thus be attributed to strong intermolecular interactions between the oppositely charged macromolecules.

Comparing the size of empty particles produced from either ALG (561 nm) or S-ALG1(1.1) (334 nm), it was found that the latter showed a significantly reduced particle size ($p < 0.0001$). Likewise, for the 10% and 20% particle systems, a significant reduction in size was observed for particles produced from S-ALG ($p < 0.005$) (Figure 3A). For the empty particles, this can be explained by the observation that the $DCC-H_2SO_4$ sulfation method produce polymers of reduced molecular weight (Ma et al. 2016). Furthermore, for the Lf loaded particles the observed relative particle size correlates with the relative binding affinity of Lf for ALG and S-ALG1(0.6). Thus, we observed ALG to show non-specific binding whereas S-ALG1(0.6) displayed strong specific binding. As a result, stronger intermolecular interactions and therefore formation of a tighter cross-linking network in the case of S-ALG1(1.1) are likely to occur.

It was observed that the addition of Ca^{2+} in the particle fabrication process when encapsulating Lf was only required when using ALG as the polymer. In this case, inconsistent results between repeat measurements were observed for two researchers when not including Ca^{2+} (Researcher 1: 475 and 608 nm, Researcher 2: 476 and 708 nm). However, when using S-ALG1(1.1) as the polymer, consistent data could be obtained. The effect of the presence of Ca^{2+} was found to be significant only for the 10% Lf particles whereas for the 20% Lf particles, no difference was observed (216 and 231 nm for particles prepared with and without Ca^{2+}, respectively) (Figure 3A). These observations correlate with the relative binding affinity of Lf for S-ALG1(0.6) compared to ALG allowing Lf to be an effective crosslinking agent in the S-ALG1(1.1) particle system. The smaller particle size for the 10% Lf particles in the presence of Ca^{2+} can be attributed to the additional crosslinking junction zones and reduced inter-chain repulsion between the carboxylate and sulfate groups of S-ALG1(1.1) (Yu et al. 2009). With increased Lf loading, the cationic protein can likely fulfill this role.

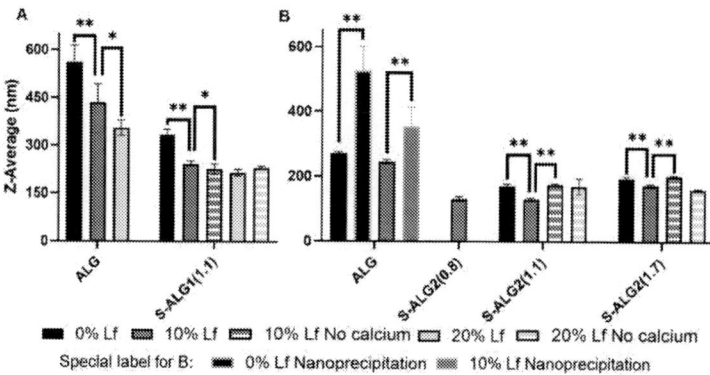

Figure 3. Particle size (z-average) of ALG and S-ALG particles. A): Series 1, all particle samples prepared by nanoprecipitation and isolated by filtration. B): Series 2, particle prepared by nanoprecipitation and sonication unless specifically labelled, and isolation by centrifugation *: $p < 0.05$, **: $p < 0.005$. For the data displayed in A, $n = 1-5$, for the data displayed in B, $n = 2$.

In addition to evaluating the size using DLS, a subset of these samples was also analyzed using NTA. Particle sizing is most commonly conducted using DLS as the method is simple. DLS measures the decay of light scattering because of the Brownian motion of particles whereas NTA allows the tagging and tracking of the particle movement through Brownian motion. The diffusion coefficient and further the hydrodynamic radius of the particle can be obtained from both methods based on Stokes-Einstein equation. For particles with a high PDI, the z-average value reported by DLS may be inaccurate, since the intensity of scattered light is proportional to the sixth power of the particle size and as such, larger particles will be over-represented (Anderson et al. 2013). Alternatively, NTA uses light scattering to identify the particles, it does not use the scattered light intensity to determine the particle size which means it is more robust against the presence of large particles/aggregates. In our study, the size of ALG particles with either 10 or 20% Lf loading were measured to be 259 ± 4 and 253 ± 5 nm, respectively, when using NTA while DLS measured sizes that were significantly larger at 361 ± 34 and 355 ± 25 nm, respectively. These observations agree with the particles displaying relatively high PDI values (up to 0.639) and the DLS measurement being based on intensity towards the larger particles. Further information regarding other technical as well as the data analysis aspects of both methods can be found in the literature (Anderson et al. 2013; Garms et al. 2021).

Second Series: Nanoprecipitation and Sonication

The second series of experiments used nanoprecipitation and sonication as the main method and centrifugation for isolation. This second series compared the polymers ALG and S-ALG2(0.8), S-ALG2(1.1), S-ALG2(1.8), the presence of Ca^{2+}, and different amounts of Lf. For this study, the particle size is reported as the z-average obtained from DLS measurements, and this size data is presented in Figure 3B. The PDI of the samples produced using nanoprecipitation and sonication were in the range of 0.178–0.300. For this series the experiments were carried out by a single researcher.

This series of experiment includes a direct comparison between the use of nanoprecipitation and sonication and the use of nanoprecipitation under stirring. This direct comparison was necessary due to the different DLS instrument used in the first and second series (Garms et al. 2021). Comparing the size of particles produced from either nanoprecipitation and sonication (271 and 246 nm for empty and 10% Lf loaded particles, respectively) or nanoprecipitation (525 and 354 nm for empty and 10% Lf loaded particles, respectively), it was found that sonication significantly reduced the particle size for ALG-based particles for both particle types ($p < 0.005$) (Figure 3B).

Ultrasonication is a high-energy method frequently used for fabrication of nano-sized particles (Table 1) due to the dispersion effects. However, it may also cause degradation of polymer chains (Dong et al. 2022) which has been evaluated for ALG in previous studies. One study indicated that the induced degradation of the ALG polymer could be attributed to C-C bond breakage by the shear forces between collapsing bubbles (Feng et al. 2017). However, Wasikiewicz et al. reported that a sufficient degradation of ALG solution was only found at a low concentration (0.25% w/w) at room temperature after 120 minutes of ultrasonication (Wasikiewicz et al. 2005). Higher degradation rate (one order of magnitude lowered M_w) was found when sonication was done at a high temperature (Feng et al. 2017; Wardhani et al. 2021). Furthermore, it has been reported that ultrasonication could change the M/G ratio of ALG and that this was affected by the ultrasonic frequency (Feng et al. 2017). Such potential changes to the ALG structure, and by inference the S-ALG structure, with regards to M_w and the M/G ratio is likely to affect the particle size.

Similar to the result of the first series, for both empty particles (271 nm, 171 nm and 195 nm for ALG, S-ALG2(1.1) and S-ALG2(1.7), respectively) and 10% Lf loaded particles (246 nm, 132 nm, 131 nm and 174 nm for ALG, S-ALG2(0.7), S-ALG2(1.1) and S-ALG2(1.7), respectively), all S-ALG samples have significantly ($p < 0.0001$) smaller size than the ALG samples (Figure 3B). When using the SO_3/py method to prepare S-ALG it has been reported that the M_w is significantly reduced (Hintze et al. 2009) and as such, for the empty particles the reduction in particle size for the S-ALG2 samples compared to ALG may be due to the reduced chain length as discussed above. As also discussed above, the reduced particle size for the S-ALG2-based particles in the presence of the Lf can be additionally attributed to strong intermolecular interactions between the oppositely charged macromolecules. There was no significant difference in particles size when comparing particles fabricated using S-ALG2(0.8) or S-ALG2(1.1). Noticeably, particles produced from S-ALG2(1.7) showed a significantly larger size ($p < 0.0001$) than other S-ALG2-based particles. The most likely explanation for the enhanced particle size for particles produced using a S-ALG2(1.7) is the enhanced electrostatic repulsion between the negatively charged groups of this polymer resulting in a more stretched polymer structure and hence larger particle size.

By comparing empty and 10% Lf particles produced from ALG, the encapsulation of Lf resulted in a significant ($p < 0.0001$) decrease in particle size, which is consistent with the trend in the first series. A similar trend could be found when comparing empty and 10% Lf loaded samples from S-ALG2(1.1) and S-ALG2(1.7) ($p < 0.0005$). In contrast to the first series, leaving out Ca^{2+} during particle fabrication resulted in significant larger particle size for 10% Lf loaded particles of the second series (Figure 3B). Common to the two series of data (Figure 3A and 3B) was that for particles without Ca^{2+}, no statistical differences are found between 10 and 20% Lf loaded particles. Furthermore, this observation was made for particles produced using different DS: particles loaded with 10% Lf (176 and 203 nm for S-ALG2(1.1) and S-ALG2(1.7), respectively), and particles loaded with 20% Lf (171 and 164 nm for S-ALG2(1.1) and S-ALG2(1.7), respectively) (Figure 3B). These observations may be due to subtle attributes from variable interactions between Lf and S-ALG as discussed above.

The zeta potential was evaluated for a number of the particle samples of the second series including blank (ALG, S-ALG2(1.1) and S-ALG2(1.7)), 10% Lf loaded (ALG, S-ALG2(0.7), S-ALG2(1.1) and S-ALG2(1.7)), and 20% loaded without Ca^{2+} (ALG, S-ALG2(1.1) and S-ALG2(1.7)). Despite the different composition of the particles in regard to the overall charge of the anionic polymer and the absence or presence of Lf, all particles displayed zeta potential values in the range of −32 to −39 mV with the most negatively charged particles being those prepared with 20% Lf. Previous studies evaluating gel particles coated with CHI have seen changes to the zeta potential approaching near neutral values and/or charge reversal (Aston et al. 2015; A.-J. Choi et al. 2011; Wang et al. 2016). The negative values obtained in the current study correlate with an outer particle surface dominated by the anionic polymer and indicate that the amount of CHI added is unable to cause charge reversal or even charge neutralisation.

Evaluation of Particle Isolation Process on Suspension Stability

Two different methods of isolating the particles were evaluated, filtration and centrifugation, for ALG particles produced with 10% Lf and Ca^{2+}. The z-average values presented in Figure 3 gives only a single size weighted average value. It is therefore important to also evaluate the particle size distribution from the DLS measurements and this can be afforded by the intensity plots shown in Figure 4 which represent one replicate for each sample type. A second replicate can be found in Figure 5 (time = 0 days). From this data it can be seen that as prepared samples fabricated by nanoprecipitation displayed a broad particle size distribution with multiple peaks (Figure 4A). Samples purified by either filtration or centrifugation showed a reduced number of large particles (around 5 μm) and although they were not eliminated by either method, they were more pronounced after filtration. After purification by either filtration or centrifugation, the main peak of the sample became narrower indicating that the sample was less polydisperse. For samples prepared by nanoprecipitation and sonication (Figure 4B), the particle size distribution is generally less broad compared to particles prepared by nanoprecipitation (Figure 4A). Purification of the particles prepared by nanoprecipitation and sonication (Figure 4B) resulted in reduced polydispersity and filtration left some large particles while centrifugation was more effective at removing the large particles.

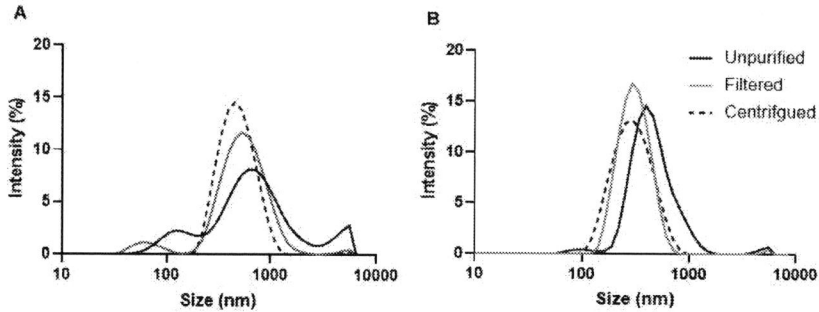

Figure 4. DLS intensity plots comparing the size distribution of particles either unpurified (solid black line); purified by filtration (solid grey line); or purified by centrifugation (broken black line). A) nanoprecipitation, B) nanoprecipitation and sonication. Samples are part of the second series and prepared using ALG with 10% Lf.

The four purified particle systems shown in Figure 4 were furthermore evaluated for their suspension stability over a 6-day period. This was done using DLS measurements and the size distribution plots are shown in Figure 5. The sample prepared by nanoprecipitation and purified by filtration showed a polydisperse size distribution and poor stability over six days suspended in water (Figure 5A). The main peak of the DLS plot of this sample shifts to a larger size over the six days (from 864 to 1048 nm). The sample prepared by nanoprecipitation and sonication and purified by filtration (Figure 5B) were less polydisperse and the main peak was found shift from 260 to 280 nm, and the width of the peak did not remain constant. However, the z-average size did not show any significant change. The instability of filtered samples may be due to the fouling of the filter by free ALG, Lf and particles (Jamal, Chang, and Zhou 2014; Gentile 2020). Fouled membranes have smaller a pore size, which may cause further fouling and thus changes to particle size and stability. It is therefore recommended not to filter this type of particel samples.

Furthermore, the type of filter, as well as the solvent system used influence the effectiveness of syringe filtration.

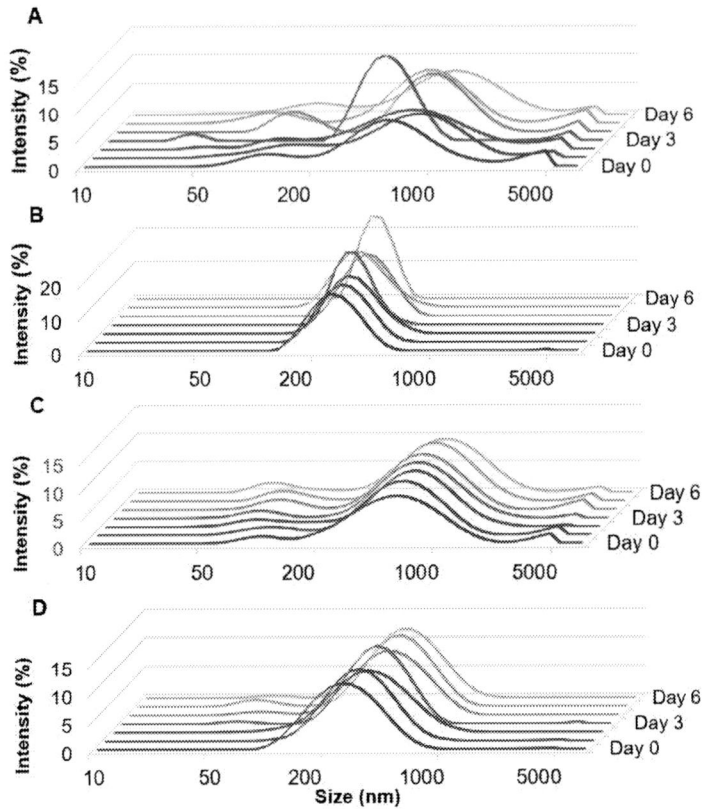

Figure 5. DLS Intensity plots of ALG-based gel particles in water measured daily over a 6-day period. A) ALG particles produced by nanoprecipitation and purified by filtration, B) ALG particles produced by nanoprecipitation and sonication and purified by filtration, C) ALG particles produced by nanoprecipitation and purified by centrifugation, D) ALG particles produced by nanoprecipitation and sonication and purified by centrifugation. All samples were based on formulation of ALG with 10% Lf.

For samples purified by centrifugation, the ones prepared by nanoprecipitation (Figure 5C) were relatively stable over the 6-day period despite the presence of large aggregates. The particle size was 450 nm and this did not change over time. Similarly, the samples prepared by nanoprecipitation and sonication (Figure 5D) were stable with the particle size remaining at 280 nm over the six days. For both sample types purified by centrifugation the zeta potential was found to be stable changing no more than 2 mV. Based on this combined data, regardless of the particle fabrication method, it is recommended that purification is done using centrifugation.

Sonication Stability of Lf

While the use of a sonic probe for the energy input was beneficial in regard to reducing the particle size and polydispersity, it is important to also evaluate if this causes any changes to the therapeutic, the Lf protein in this case. The Lf stability was therefore evaluated using CD spectroscopy which can reflect the secondary and tertiary structural changes of proteins (Yang, Wu, and Martinez 1986).

The CD spectra of Lf presented in Figure 6 are similar to those previously reported (Bokkhim et al. 2013). Negative ellipticity values are observed from 208 to 250 nm. Both the heat-treated Lf (Lf-H) and sonicated Lf (Lf-S) samples display similar shape but lower intensity than Lf (Figure 6A). This indicates that the secondary structure was largely maintained after heat or sonication treatment. In a similar manner, the spectra in the 250 to 300 nm range (Figure 6B) displayed only minor changes in intensity for the treated Lf samples compared to that of untreated Lf which indicates that the tertiary structure is not adversely altered.

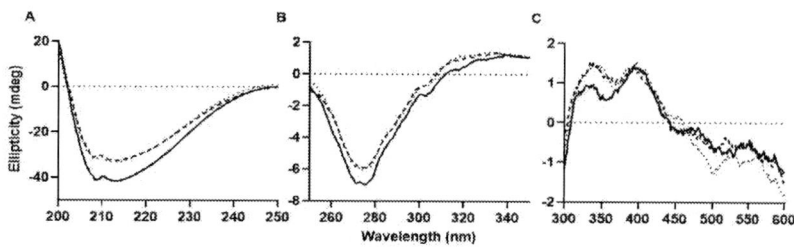

Figure 6. CD spectra of Lf (—); Lf heat treated (20 minutes of 55 °C) (Lf-H, ---); Lf sonication treated (20 minutes of 25 W probe-type sonication) (Lf-S, ⋯). A) wavelength region 200–250 nm, B) wavelength region 250–300 nm, C) wavelength region 300–600 nm.

In the 300–600 nm region (Figure 6C), the ellipticities of Lf-S and Lf-H are somewhat different to the original Lf sample. The peak around 350 nm has higher ellipticity for Lf-S and LF-H compared to Lf, while at wavelengths above 450 nm, only the spectrum of Lf-S displays deviation from the Lf spectrum. While these observations may be attributed to conformational differences of the native and treated Lf samples, it should be noted that larger spectral differences were observed for Lf with different iron saturation levels (Bokkhim et al. 2013).

Our observations agree with that reported in the literature. The heat stability of native Lf was evaluated using DSC and a denaturation peak with a T_{max} value of 61.3 °C and an onset temperature above 50 °C was determined (Bokkhim et al. 2013). No study found evidence of significant denaturation when heating a Lf sample at 55 °C. In addition, Steijns et al. reported that native bovine Lf remained 75% of its structural integrity after heat treatment at 72 °C for 8 minutes (Steijns and Van Hooijdonk 2000). It is worth mentioning that holo-Lf (fully iron saturated), was found to have much higher thermal stability (T_{max} > 90 °C) (Abe et al. 1991; Bokkhim et al. 2013; Wakabayashi, Yamauchi, and Takase 2006). In regard to sonication stability, Lf was found to retain 96% in its structure after 15 minutes of sonication at 600 W (Miao et al. 2020). Overall, it appears safe to use sonication when producing Lf containing gel particles and if higher protein stability is required, converting native Lf to holo-Lf would be a suitable strategy.

Conclusion

This study has extended previous research into the use of nanoprecipitation alone and nanoprecipitation in combination with sonication for the production of ALG and S-ALG-based gel particles. Using SPR, the polymer S-ALG was found to have high affinity to Lf ($2.0 \times 10^7 \, M^{-1}$) while ALG only showed non-specific binding. These observations correlated with the smaller particle sizes obtained when using S-ALG for Lf encapsulation, although, the reduced molecular mass of S-ALG (regardless of synthesis method) may also play a part. Calcium ions for ionic cross linking was only required for Lf encapsulation in ALG-based gel particles while gel particles fabricated using S-ALG could omit this step as Lf acted as a crosslinking agent. It was found that the use of sonication in combination with nanoprecipitation reduced the particle size as well as the particle size distribution. Purification of the gel particles was evaluated using either filtration or centrifugation with the latter yielding smaller particles with lower dispersity. Furthermore, the particle suspension stability over six days was enhanced when using centrifugation. The structural integrity of Lf as assessed by CD spectroscopy was very good and similar to that of a 20 minutes 55°C heat treatment. This suggests that the use of sonication in combination with nanoprecipitation is a viable method for encapsulation of the therapeutic protein Lf.

Acknowledgments

The authors thank Dr Amanda Nouwens for the feedback, technical support as well as insightful discussion for the Biacore™ instrument usage, data collection and analysis. The authors are thankful to Ms Alexandra Mutch and Mr Desmond Sim for donating some of the sulfated alginate samples. Author AA is thankful to The University of Queensland, Australia (UQ Postdoctoral Research fellowships; 2015-2016), National Health and Medical Research Council, Australia for the Early Career Research Fellowship (APP1123340, 2017-2021) as well as the UQ Amplify Fellowship (2021 Dec-2023 Nov) for providing the financial support. Author Mr JY acknowledges the scholarship from The University of Queensland and China Scholarship Council (CSC). Author BD acknowledges the Australian National University (ANU) for allowing to work at the University of Queensland during a summer research project. Author AG acknowledge l'Ecole Nationale Supérieure de Chimie de Rennes (the National School of Chemistry of Rennes) for allowing her to do her internship at the University of Queensland and she is thankful to the University of Queensland for letting her be part of this project. Author YG acknowledges the funding and support from the CSC.

References

Abe, Hiroaki, Hitoshi Saito, Hiroshi Miyakawa, Yoshitaka Tamura, Seiichi Shimamura, Eiji Nagao, and Mamoru Tomita. 1991. "Heat stability of bovine lactoferrin at acidic pH." *Journal of dairy science* 74 (1): 65-71.

Abied, H, A Brulet, and JM Guenet. 1990. "Physical gels from PVC: molecular structure of pregels and gels by low-angle neutron scattering." *Colloid and Polymer Science* 268 (5): 403-413.

Anderson, Will, Darby Kozak, Victoria A Coleman, Åsa K Jämting, and Matt Trau. 2013. "A comparative study of submicron particle sizing platforms: accuracy, precision and resolution analysis of polydisperse particle size distributions." *Journal of colloid and interface science* 405: 322-330.

Arlov, Øystein, and Gudmund Skjåk-Bræk. 2017. "Sulfated alginates as heparin analogues: a review of chemical and functional properties." *Molecules* 22 (5): 778.

Aston, Robyn, Medini Wimalaratne, Aidan Brock, Gwendolyn Lawrie, and Lisbeth Grøndahl. 2015. "Interactions between chitosan and alginate dialdehyde biopolymers and their layer-by-layer assemblies." *Biomacromolecules* 16 (6): 1807-1817.

Basha, S Khaleel, M Syed Muzammil, R Dhandayuthabani, and V Sugantha Kumari. 2021. "Development of nanoemulsion of Alginate/Aloe vera for oral delivery of insulin." *Materials Today: Proceedings* 36: 357-363.

Bokkhim, Huma, Nidhi Bansal, Lisbeth GrØndahl, and Bhesh Bhandari. 2013. "Physico-chemical properties of different forms of bovine lactoferrin." *Food Chemistry* 141 (3): 3007-3013.

———. 2016a. "Characterization of alginate–lactoferrin beads prepared by extrusion gelation method." *Food hydrocolloids* 53: 270-276.

———. 2016b. "In-vitro digestion of different forms of bovine lactoferrin encapsulated in alginate micro-gel particles." *Food Hydrocolloids* 52: 231-242.

Bokkhim, Huma, Trang Tran, Nidhi Bansal, Lisbeth Grøndahl, and Bhesh Bhandari. 2014. "Evaluation of different methods for determination of the iron saturation level in bovine lactoferrin." *Food chemistry* 152: 121-127.

Bourganis, Vassilis, Theodora Karamanidou, Olga Kammona, and Costas Kiparissides. 2017. "Polyelectrolyte complexes as prospective carriers for the oral delivery of protein therapeutics." *European Journal of Pharmaceutics and Biopharmaceutics* 111: 44-60.

Chen, Zhuoxin, Peng Yu, Zhangshu Miao, Haochen Zhang, Hong Xiao, Jing Xie, Chunmei Ding, and Jianshu Li. 2021. "Sulfated alginate based complex for sustained calcitonin delivery and enhanced osteogenesis." *Biomedical Materials* 16 (3): 035022.

Choi, Ae-Jin, Chul-Jin Kim, Yong-Jin Cho, Jae-Kwan Hwang, and Chong-Tai Kim. 2011. "Characterization of capsaicin-loaded nanoemulsions stabilized with alginate and chitosan by self-assembly." *Food and Bioprocess Technology* 4 (6): 1119-1126.

Choi, Charles J, Ian D Block, Brian Bole, David Dralle, and Brian T Cunningham. 2009. "Label-free photonic crystal biosensor integrated microfluidic chip for determination of kinetic reaction rate constants." *IEEE Sensors Journal* 9 (12): 1697-1704.

Choi, Hong Joon, Somang Choi, Jae Gyoon Kim, Mi Hyun Song, Kyu-Sik Shim, Youn-Mook Lim, Hak-Jun Kim, Kyeongsoon Park, and Sung Eun Kim. 2020. "Enhanced tendon restoration effects of anti-inflammatory, lactoferrin-immobilized, heparin-polymeric nanoparticles in an Achilles tendinitis rat model." *Carbohydrate polymers* 241: 116284.

Choi, Yo Han, Chang Beom Kim, Hyo Bong Hong, and Woon Seob Lee. 2017. "Production of Alginate Sub-micron Particles and Their Biological Application." *Bulletin of the Korean Chemical Society* 38 (4): 507-510.

Dong, Jun, Zonghua Wang, Fangfang Yang, Huiqi Wang, Xuejun Cui, and Zhanfeng Li. 2022. "Update of ultrasound-assembling fabrication and biomedical applications for heterogeneous polymer composites." *Advances in Colloid and Interface Science*: 102683.

Duchardt, Falk, Ivo R Ruttekolk, Wouter PR Verdurmen, Hugues Lortat-Jacob, Jochen Bürck, Hansjörg Hufnagel, Rainer Fischer, Maaike Van den Heuvel, Dennis WPM Löwik, and Geerten W Vuister. 2009. "A cell-penetrating peptide derived from human lactoferrin with conformation-dependent uptake efficiency." *Journal of Biological Chemistry* 284 (52): 36099-36108.

Feng, Liping, Yanping Cao, Duoxia Xu, Shaojia Wang, and Jie Zhang. 2017. "Molecular weight distribution, rheological property and structural changes of sodium alginate induced by ultrasound." *Ultrasonics sonochemistry* 34: 609-615.

Freeman, Inbar, and Smadar Cohen. 2009. "The influence of the sequential delivery of angiogenic factors from affinity-binding alginate scaffolds on vascularization." *Biomaterials* 30 (11): 2122-2131.

Freeman, Inbar, Alon Kedem, and Smadar Cohen. 2008. "The effect of sulfation of alginate hydrogels on the specific binding and controlled release of heparin-binding proteins." *Biomaterials* 29 (22): 3260-3268.

Garms, Bruna C, Hamish Poli, Darcy Baggley, Felicity Y Han, Andrew K Whittaker, A Anitha, and Lisbeth Grøndahl. 2021. "Evaluating the effect of synthesis, isolation, and characterisation variables on reported particle size and dispersity of drug loaded PLGA nanoparticles." *Materials Advances* 2 (17): 5657-5671.

Gentile, Luigi. 2020. "Protein–polysaccharide interactions and aggregates in food formulations." *Current Opinion in Colloid & Interface Science* 48: 18-27.

Grøndahl, Lisbeth, Gwendolyn Lawrie, A Anitha, and Aparna Shejwalkar. 2020. "Applications of alginate biopolymer in drug delivery." In *Biointegration of medical implant materials*, 375-403. Elsevier.

Hadiya, Safy, Radwa Radwan, Menna Zakaria, Tahra El-Sherif, Mostafa A Hamad, and Mahmoud Elsabahy. 2021. "Nanoparticles integrating natural and synthetic polymers for in vivo insulin delivery." *Pharmaceutical development and technology* 26 (1): 30-40.

Hintze, Vera, Stephanie Moeller, Matthias Schnabelrauch, Susanne Bierbaum, Manuela Viola, Hartmut Worch, and Dieter Scharnweber. 2009. "Modifications of hyaluronan influence the interaction with human bone morphogenetic protein-4 (hBMP-4)." *Biomacromolecules* 10 (12): 3290-3297.

Hornyak, Gabor L, Harry F Tibbals, Joydeep Dutta, and John J Moore. 2008. *Introduction to nanoscience and nanotechnology*. CRC press.

Jamal, Sondus, Sheng Chang, and Hongde Zhou. 2014. "Filtration behaviour and fouling mechanisms of polysaccharides." *Membranes* 4 (3): 319-332.

Kamei, Kaeko, Xiaofeng Wu, Xinyan Xu, Kazuhiro Minami, Nguyen Tien Huy, Ryo Takano, Hisao Kato, and Saburo Hara. 2001. "The analysis of heparin–protein interactions using evanescent wave biosensor with regioselectively desulfated heparins as the ligands." *Analytical Biochemistry* 295 (2): 203-213.

Kordbacheh, Emad, Shahram Nazarian, Abbas Hajizadeh, and Davood Sadeghi. 2018. "Entrapment of LTB protein in alginate nanoparticles protects against Enterotoxigenic Escherichia coli." *Apmis* 126 (4): 320-328.

Kumar, Vijay, Vikash Kumar Yadav, Imtaiyaz Hassan, Abhay Kumar Singh, Sharmistha Dey, Sarman Singh, Tej P Singh, and Savita Yadav. 2012. "Kinetic and structural studies on the interactions of heparin and proteins of human seminal plasma using surface plasmon resonance." *Protein and Peptide Letters* 19 (8): 795-803.

Lawrie, Gwen, Imelda Keen, Barry Drew, Adrienne Chandler-Temple, Llewellyn Rintoul, Peter Fredericks, and Lisbeth Grøndahl. 2007. "Interactions between alginate and chitosan biopolymers characterized using FTIR and XPS." *Biomacromolecules* 8 (8): 2533-2541.

Lee, Kuen Yong, and David J Mooney. 2012. "Alginate: properties and biomedical applications." *Progress in polymer science* 37 (1): 106-126.

Legrand, Dominique, Annick Pierce, Elisabeth Elass, Mathieu Carpentier, Christophe Mariller, and Joel Mazurier. 2008. "Lactoferrin structure and functions." *Bioactive components of milk*: 163-194.

Liao, I-Chien, Andrew CA Wan, Evelyn KF Yim, and Kam W Leong. 2005. "Controlled release from fibers of polyelectrolyte complexes." *Journal of Controlled Release* 104 (2): 347-358.

Ma, Lang, Chong Cheng, Chuanxiong Nie, Chao He, Jie Deng, Lingren Wang, Yi Xia, and Changsheng Zhao. 2016. "Anticoagulant sodium alginate sulfates and their mussel-inspired heparin-mimetic coatings." *Journal of Materials Chemistry B* 4 (19): 3203-3215.

Maatouk, Batoul, Miran A Jaffa, Mia Karam, Duaa Fahs, Wared Nour-Eldine, Anwarul Hasan, Ayad A Jaffa, and Rami Mhanna. 2021. "Sulfated alginate/polycaprolactone double-emulsion nanoparticles for enhanced delivery of heparin-binding growth factors in wound healing applications." *Colloids and Surfaces B: Biointerfaces* 208: 112105.

Mansourpour, Maryam, Reza Mahjub, Mohsen Amini, Seyed Naser Ostad, Elnaz Sadat Shamsa, Morteza Rafiee-Tehrani, and Farid Abedin Dorkoosh. 2015. "Development of acid-resistant alginate/trimethyl chitosan nanoparticles containing cationic β-cyclodextrin polymers for insulin oral delivery." *Aaps Pharmscitech* 16 (4): 952-962.

Martınez-Gomis, J, A Fernández-Solanas, M Vinas, P Gonzalez, ME Planas, and S Sanchez. 1999. "Effects of topical application of free and liposome-encapsulated lactoferrin and lactoperoxidase on oral microbiota and dental caries in rats." *Archives of oral biology* 44 (11): 901-906.

Mhanna, Rami, Jana Becher, Matthias Schnabelrauch, Rui L Reis, and Iva Pashkuleva. 2017. "Sulfated alginate as a mimic of sulfated glycosaminoglycans: binding of growth factors and effect on stem cell behavior." *Advanced Biosystems* 1 (7): 1700043.

Mhanna, Rami, Aditya Kashyap, Gemma Palazzolo, Queralt Vallmajo-Martin, Jana Becher, Stephanie Möller, Matthias Schnabelrauch, and Marcy Zenobi-Wong. 2014. "Chondrocyte culture in three dimensional alginate sulfate hydrogels promotes proliferation while maintaining expression of chondrogenic markers." *Tissue Engineering Part A* 20 (9-10): 1454-1464.

Miao, Wanlu, Ru He, Li Feng, Kai Ma, Changliang Zhang, Jianzhong Zhou, Xiaohong Chen, Xin Rui, Qiuqin Zhang, and Mingsheng Dong. 2020. "Study on processing stability and fermentation characteristics of donkey milk." *LWT* 124: 109151.

Mukhopadhyay, Piyasi, Subhajit Maity, Sudipto Mandal, Abhay Sankar Chakraborti, AK Prajapati, and Patit Paban Kundu. 2018. "Preparation, characterization and in vivo evaluation of pH sensitive, safe quercetin-succinylated chitosan-alginate core-shell-corona nanoparticle for diabetes treatment." *Carbohydrate polymers* 182: 42-51.

Nakagawa, Keiko, Kosuke Nakamura, Yuji Haishima, Makiko Yamagami, Kana Saito, Hiromi Sakagami, and Haruko Ogawa. 2009. "Pseudoproteoglycan (pseudoPG) probes that simulate PG macromolecular structure for screening and isolation of PG-binding proteins." *Glycoconjugate journal* 26 (8): 1007-1017.

Osmond, Ronald IW, Warren C Kett, Spencer E Skett, and Deirdre R Coombe. 2002. "Protein–heparin interactions measured by BIAcore 2000 are affected by the method of heparin immobilization." *Analytical biochemistry* 310 (2): 199-207.

Pawar, Siddhesh N, and Kevin J Edgar. 2012. "Alginate derivatization: A review of chemistry, properties and applications." *Biomaterials* 33 (11): 3279-3305.

Polyak, Boris, Shimona Geresh, and Robert S Marks. 2004. "Synthesis and characterization of a biotin-alginate conjugate and its application in a biosensor construction." *Biomacromolecules* 5 (2): 389-396.

Rajaonarivony, M, C Vauthier, G Couarraze, F Puisieux, and P Couvreur. 1993. "Development of a new drug carrier made from alginate." *Journal of pharmaceutical sciences* 82 (9): 912-917.

Re'em, Tali, and Smadar Cohen. 2011. "Microenvironment design for stem cell fate determination." *Tissue engineering III: cell-surface interactions for tissue culture*: 227-262.

Roger, Yvonne, Steffen Sydow, Laura Burmeister, Henning Menzel, and Andrea Hoffmann. 2020. "Sustained release of TGF-β3 from polysaccharide nanoparticles induces chondrogenic differentiation of human mesenchymal stromal cells." *Colloids and Surfaces B: Biointerfaces* 189: 110843.

Ruvinov, Emil, Inbar Freeman, Roei Fredo, and Smadar Cohen. 2016. "Spontaneous coassembly of biologically active nanoparticles via affinity binding of heparin-binding proteins to alginate-sulfate." *Nano Letters* 16 (2): 883-888.

Samarasinghe, Rasika M, Rupinder K Kanwar, and Jagat R Kanwar. 2014. "The effect of oral administration of iron saturated-bovine lactoferrin encapsulated chitosan-nanocarriers on osteoarthritis." *Biomaterials* 35 (26): 7522-7534.

Sarmento, Bruno, Susana Martins, Antonio Ribeiro, Francisco Veiga, Ronald Neufeld, and Domingos Ferreira. 2006. "Development and comparison of different nanoparticulate polyelectrolyte complexes as insulin carriers." *International Journal of Peptide Research and Therapeutics* 12 (2): 131-138.

Saxena, Abhishek, Arpita Bhattacharya, Satish Kumar, Irving R Epstein, and Rachana Sahney. 2017. "Biopolymer matrix for nano-encapsulation of urease–A model protein and its application in urea detection." *Journal of colloid and interface science* 490: 452-461.

Schmidt, John, Min Kyung Lee, Eunkyung Ko, Jae Hyun Jeong, Luisa A DiPietro, and Hyunjoon Kong. 2016. "Alginate sulfates mitigate binding kinetics of proangiogenic growth factors with receptors toward revascularization." *Molecular pharmaceutics* 13 (7): 2148-2154.

Shamekhi, Fatemeh, Elnaz Tamjid, and Khosro Khajeh. 2018. "Development of chitosan coated calcium-alginate nanocapsules for oral delivery of liraglutide to diabetic patients." *International journal of biological macromolecules* 120: 460-467.

Sohail, Rabia, and Shah Rukh Abbas. 2020. "Evaluation of amygdalin-loaded alginate-chitosan nanoparticles as biocompatible drug delivery carriers for anticancerous efficacy." *International journal of biological macromolecules* 153: 36-45.

Steijns, Jan M, and ACM Van Hooijdonk. 2000. "Occurrence, structure, biochemical properties and technological characteristics of lactoferrin." *British Journal of Nutrition* 84 (S1): 11-17.

Tammam, Salma N, Hassan ME Azzazy, and Alf Lamprecht. 2018. "Nuclear and cytoplasmic delivery of lactoferrin in glioma using chitosan nanoparticles: Cellular location dependent-action of lactoferrin." *European Journal of Pharmaceutics and Biopharmaceutics* 129: 74-79.

Wakabayashi, Hiroyuki, Koji Yamauchi, and Mitsunori Takase. 2006. "Lactoferrin research, technology and applications." *International Dairy Journal* 16 (11): 1241-1251.

Wang, Fang, Siqian Yang, Jian Yuan, Qinwei Gao, and Chaobo Huang. 2016. "Effective method of chitosan-coated alginate nanoparticles for target drug delivery applications." *Journal of biomaterials applications* 31 (1): 3-12.

Wardhani, Dyah Hesti, Nita Aryanti, Abdul Aziz, Rinda Ameliya Firdhaus, and Hana Nikma Ulya. 2021. "Ultrasonic degradation of alginate: A matrix for iron encapsulation using gelation." *Food Bioscience* 41: 100803.

Wasikiewicz, Jaroslaw M, Fumio Yoshii, Naotsugu Nagasawa, Radoslaw A Wach, and Hiroshi Mitomo. 2005. "Degradation of chitosan and sodium alginate by gamma radiation, sonochemical and ultraviolet methods." *Radiation Physics and Chemistry* 73 (5): 287-295.

Xiong, Sha, Zhongjun Li, Yao Liu, Qun Wang, Jingshan Luo, Xiaojia Chen, Zhongjian Xie, Yuan Zhang, Han Zhang, and Tongkai Chen. 2020. "Brain-targeted delivery shuttled by black phosphorus nanostructure to treat Parkinson's disease." *Biomaterials* 260: 120339.

Yang, Jen Tsi, Chuen-Shang C Wu, and Hugo M Martinez. 1986. "[11] Calculation of protein conformation from circular dichroism." In *Methods in enzymology*, 208-269. Elsevier.

Yoncheva, Krassimira, Maria Merino, Aslihan Shenol, Nikolay T Daskalov, Petko St Petkov, Georgi N Vayssilov, and Maria J Garrido. 2019. "Optimization and in-vitro/in-vivo evaluation of doxorubicin-loaded chitosan-alginate nanoparticles using a melanoma mouse model." *International journal of pharmaceutics* 556: 1-8.

Yu, Cui-Yun, Hua Wei, Qiao Zhang, Xian-Zheng Zhang, Si-Xue Cheng, and Ren-Xi Zhuo. 2009. "Effect of ions on the aggregation behavior of natural polymer alginate." *The Journal of Physical Chemistry B* 113 (45): 14839-14843.

Zimet, Patricia, Álvaro W Mombrú, Ricardo Faccio, Giannina Brugnini, Iris Miraballes, Caterina Rufo, and Helena Pardo. 2018. "Optimization and characterization of nisin-loaded alginate-chitosan nanoparticles with antimicrobial activity in lean beef." *LWT* 91: 107-116.

Biographical Sketches

A. Anitha, PhD

Affiliation: School of Chemistry and Molecular Biosciences, The University of Queensland, Australia

Education: Doctor of Philosophy

Business Address: Chemistry building, The University of Queensland, St Lucia, QLD 4072, Australia

Research and Professional Experience: Dr A. Anitha holds a PhD in nano-biotechnology (2014) and held two prestigious post-doctoral positions (NHMRC Early Career Research Fellowship and UQ Post-Doctoral Fellowship) at the University of Queensland. Presently she holds a UQ Amplify Lecturer position (Research and teaching position) within the School of Chemistry and Molecular Bioscience, The University of Queensland. Her research focuses on the development of natural and synthetic polymer-based therapeutic formulations towards drug delivery especially towards cancer as well as bone fracture healing applications. Her skills and expertise allow her to work across the breath of discipline areas that covers biomaterials science (including chemistry), cell biology and in vivo studies in small animal models. More recently, Dr Sudheesh Kumar has extended her research expertise to the field of minimum intervention dentistry and awarded a GC Australasia Dental PTY LTD Minimum Intervention Dentistry Research Award (2020).

Professional Appointments: UQ Amplify Lecturer

Honors:

2005	2nd Rank Holder (BSc Polymer Chemistry), Kannur University, Kerala, India
2007	76 Percentile in the Graduate Aptitude Test in Engineering; 2007 from Kerala, India.
2010	Senior Research Fellowship from the Council of Scientific and Industrial Research; India (3 Years)
2010	Best Poster Award in Third Bangalore Nano International Conference, India.
2010	Selected as young researcher for an expense free participation in an international Conference (International Conference on Nanoscience and Technology (ICONSAT), India.
2015	UQ Post-Doctoral Fellowship from the University of Queensland for the duration of 3 years (2015-2017), Relinquished on December 31st 2016.

2015	ASBTE Travel Award (Given a talk in the 5th International Symposium of Surface and Interface of Biomaterials & 24th Annual Conference of the Australasian Society for Biomaterials and Tissue Engineering, Sydney, Australia).
2016	Early Career Researchers (ECR) Travel Award and ASBTE Travel Award (presented a poster in the 10th World Biomaterials Congress, May 2016, Montreal, QC Canada).
2017	National Health and Medical Research Council (NHMRC) Early Career Fellowships (Peter Doherty Biomedical Fellowship) for the duration of 4 years (2017-2021).
2018	UQ Early Career Researcher, Research Donation Generic Grant (2018-2019)

Publications from the Last 3 Years:

Book Chapters:
1. Grøndahl, L, G Lawrie, A Anitha, A Shejwalkar (2020) Chapter 9. Applications of alginate biopolymer in drug delivery. *Biointegration of Medical Implant Materials* 2nd Edition. (Woodhead Publishing Series in Biomaterials).

   Peer-Reviewed Journal Articles:
2. Garms, BC, H Poli, D Baggley, FY Han, AK Whittaker, A Anitha[#], L Grøndahl (2021), Evaluating the effect of synthesis, isolation, and characterisation variables on reported particle size and dispersity of drug loaded PLGA nanoparticles, *Materials Advances,* 2021, 2, 5657-5671. # Co-Corresponding author.
3. Poli, H, A Mushtaq, CA Bell, RP McGeary, L Grøndahl, A Anitha*(2021) Optimisation of alendronate conjugation to polyethylene glycol for functionalisation of biopolymers and nanoparticles, *European Polymer Journal,* 155, 2021, 110571 10.1016/j.eurpolymj.2021.110571 (Journal Impact Factor= 4.6) * Corresponding author.
4. Mushtaq, A, L Li, A Anitha[#], L Grondahl (2021), Chitosan nanomedicine in cancer therapy: targeted delivery and cellular uptake, *Macromolecular Bioscience*, 2021, 21(5): e2100005. doi: 10.1002/mabi.202100005. (Journal Impact Factor: 4.979). [#] Co-Corresponding Author.

5. Anitha, A, NL Fletcher, ZH Houston, KJ Thurecht, L Grøndahl (2021), Evaluation of the In vivo fate of ultrapure alginate in a BALB/c mice model, *Carbohydrate Polymers*, 2021, 262, 117947 (Journal Impact Factor: 9.38).
6. Poli, H, AL Mutch, A Anitha, S Ivanovski, C Vaquette, DG Castner, MN Gómez-Cerezo, L Grøndahl (2020). Evaluation of surface layer stability of surface-modified polyester biomaterials. *Biointerphases*. 2020, 15, 061010. doi: 10.1116/6.0000687. (Journal Impact Factor: 2.456)
7. Mohd Hidzir, N*, A Anitha*, K Kępa, DJT Hill, L Jorgensen, L Grøndahl (2020). Protein adsorption to poly (tetrafluoroethylene) membranes modified with grafted poly (acrylic acid) chains. *Biointerphases* 15, 031011 (*Journal Impact Factor:* 2.456) *Equal first authors.

Full publication list: ORCID 0000-0003-0964-741X.

Lisbeth Grøndahl, PhD

Affiliation: School of Chemistry and Molecular Biosciences, The University of Queensland, Australia

Education: Doctor of Philosophy

Business Address: Chemistry building, The University of Queensland, St Lucia, QLD 4072, Australia

Research and Professional Experience: Professor Grøndahl obtained her PhD from the University of Copenhagen (1995), held various Fellowships and academic positions and was appointed a lecturer in Materials Chemistry at the University of Queensland in 2002. She works in the interdisciplinary field of bio-interface science with a focus on development and evaluation of approaches to create functional polymeric biomaterials in the form of membranes, scaffolds, hydrogels and nanoparticles. She leads a research group which currently includes 6 PhD students and a UQ Amplify Fellow. Professor Grøndahl was in 2016 internationally recognized for her research standing and awarded the title of Fellow of Biomaterials Science and Engineering. Recently she has been awarded the STRAUMANN GROUP

RESEARCH AWARD (2019) and the GC Australasia Dental PTY LTD Minimum Intervention Dentistry Research Award (2020).

Professional Appointments: Professor

Honors:

2011	International Year of Chemistry Jim O'Donnell Lecturer sponsored by the Queensland Education node of RACI
2014	UQ Promoting Women Fellowship
2016	International honor: Fellow of Biomaterials Science and Engineering – FBSE
2018	Highlighted as international expert: Selected as one of 15 Women in Biointerface Science by the American Vacuum Society journal Biointerphases
2016 – 2023	Affiliate Principal Research Fellow in the Australasian Institute for Bioengineering and Nanotechnology
2019	Award of Excellence, The Australasian Society for Biomaterials and Tissue Engineering
2019	School of Chemistry and Molecular Biosciences (UQ) HDR Supervision Excellence Award

Publications from the Last 3 Years:

Book Chapters:
1. Wentrup-Byrne, E, L Grøndahl, A Chandler-Temple, Chapter 3. Replacement materials for facial reconstruction at the soft tissue - bone interface, in *Biointegration of medical implant materials: science and design,* Edited by Dr. Chandra P. Sharma, Woodhead Publishing Limited, 2nd Edition, 2020.
2. Grøndahl, L, G Lawrie, A Anitha, A Jejurikar; Chapter 9 "Applications of Alginate Bioipolymer in Drug Delivery" in *"Biointegration of medical implant materials: science and design"* Edited by Dr. Chandra P. Sharma, Woodhead Publishing Limited, 2nd Edition, 2020.

Peer-Reviewed Journal Articles:
3. Gomez-Cerezo, MN, R Patel, C Vaquette, L Grøndahl, M Lu, In vitro evaluation of porous poly(hydroxybutyrate-co-hydroxyvalerate)/akermanite composite scaffolds manufactured using selective laser sintering, *Biomaterials Advances*, https://doi.org/10.1016/j.bioadv.2022.212748.
4. Cesar, MB, H Poli, RD Piazza, RFC Marques, RD Herculano, L Grøndahl, Dispersion of hydroxyapatite nanoparticles in natural rubber latex and poly lactic acid based matrices, *J. Appl. Polym. Sci.* (2022) e52165.
5. Bryceson, KP, S Leigh, S Sarwar, L Grøndahl, Affluent Effluent: Visualizing the invisible during the development of an algal bloom using systems dynamics modelling and augmented reality technology, *Environmental Modelling & Software* 147 (2022) 105253.
6. Garms, BC, H Poli, D Baggley, FY Han, AK Whittaker, A Anitha, L Grøndahl (2021), Evaluating the effect of synthesis, isolation, and characterisation variables on reported particle size and dispersity of drug loaded PLGA nanoparticles, *Materials Advances*, 2021, 2, 5657-5671.
7. Poli, H, A Mushtaq, CA Bell, RP McGeary, L Grøndahl, A Anitha (2021) Optimisation of alendronate conjugation to polyethylene glycol for functionalisation of biopolymers and nanoparticles, *European Polymer Journal*, 155, 2021, 110571.
8. Jayawardena, I, K Wilson, M Plebanski, L Grøndahl, S Corrie, Morphology and Composition of Immunodiffusion Precipitin Complexes Evaluated via Microscopy and Proteomics, *Journal of Proteome Research* (2021) 20 (5), 2618-2627.
9. Kępa, K, N Amiralian, DJ Martin, L Grøndahl, Grafting from cellulose nanofibres with naturally-derived oil to reduce water absorption, *Polymer* (2021) 222, 123659.
10. Mushtaq, A, L Li, A Anitha, L Grøndahl (2021), Chitosan nanomedicine in cancer therapy: targeted delivery and cellular uptake, *Macromolecular Bioscience*, 2021, 21(5): e2100005.
11. Anitha, A, NL Fletcher, ZH Houston, KJ Thurecht, L Grøndahl (2021), Evaluation of the In vivo fate of ultrapure alginate in a BALB/c mice model, *Carbohydrate Polymers*, 2021, 262, 117947.
12. Patel, R, D Monticone, M Lu, L Grøndahl, H Huang, Hydrolytic degradation of porous PHBV scaffolds manufactured using selective

laser sintering, *Polymer Degradation and Stability* 187 (2021) 109545.
13. Morise, BT, AL Mutch, BC Garms, R Donizetti Herculano, L Grøndahl, Evaluation of acrylic acid grafting on the loading and release of Scopolamine Butylbromide from polymeric matrices for future Sialorrhea treatment, *J Appl Polym Science* 2021; 138(13): e50117.
14. Poli, H, AL Mutch, A Anitha, S Ivanovski, C Vaquette, DG Castner, MN Gómez-Cerezo, L Grøndahl (2020). Evaluation of surface layer stability of surface-modified polyester biomaterials. *Biointerphases*. 2020, 15, 061010.
15. Cao, P, FY Han, L Grøndahl, ZP Xu, L Li, Enhanced oral vaccine efficacy of polysaccharide-coated calcium phosphate nanoparticles, *ACS Omega* (2020) 5 (29), 18185-18197.
16. Mohd Hidzir, N, A Anitha, K Kępa, DJT Hill, L Jorgensen, L Grøndahl (2020). Protein adsorption to poly(tetrafluoroethylene) membranes modified with grafted poly(acrylic acid) chains. *Biointerphases* 15, 031011.
17. Kępa, K, CM Chaléat, N Amiralian, L Grøndahl, W Batchelor, DJ Martin. Properties and specific energy consumption of spinifex-derived lignocellulose fibers produced using different mechanical processes. *Cellulose,* (2019) 26:6555–6569.
18. Diermann, SH, M Lu, M Dargusch, L Grøndahl, H Huang, Akermanite reinforced PHBV scaffolds manufactured using selective laser sintering, *Journal of Biomedical Materials Research Part B: Applied Biomaterials*, (2019) DOI: 10.1002/jbm.b.34349.

Full publication list: ORCID 0000-0001-6012-9808.

Chapter 3

An Insight into Properties and Applications of Alginate

Abhishek Saxena and Archana Tiwari[*]

Diatom Research Laboratory, Amity Institute of Biotechnology, Amity University, Noida, Uttar Pradesh, India

Abstract

Alginates, which are natural multifunctional polymers, have gained popularity in the biomedical and pharmaceutical industries in recent decades due to their unusual physicochemical features and diverse biological activities. Film-forming ability, pH responsiveness, gelling, hydrophilicity, biocompatibility, biodegradability, non-toxic, processability, and ionic crosslinking are only a few of the features of alginates. Food, pharmaceuticals, dental uses, welding rods, and scaffolding are just a few of the commercial applications of alginate. The cosmetics and healthcare sectors have demonstrated a significant deal of interest in biodegradable polymers in general, and alginates in particular, during the last few decades due to their gelling and non-toxic qualities, as well as their abundance in nature. The purpose of this chapter is to describe alginates' unique properties, as well as to examine their current and potential applications.

Keywords: alginate, egg box model, biocompatible, pharmaceuticals, α-L-guluronic acid, β-D-mannuronic acid

---

[*] Corresponding Author's Email: panarchana@gmail.com.

In: Properties and Applications of Alginate
Editor: Michael Y. Wilkerson
ISBN: 979-8-88697-371-6
© 2022 Nova Science Publishers, Inc.

Introduction

Polymers are chemical molecules that are important in the advancement of medicine and engineering in everyday life. Polymers are made up of multiple monomer repeating units that determine the structure and properties. A variety of polymeric materials are being used in medicine. However, biodegradable polymers are gaining popularity (Reddy, 2021). Natural polymers are materials that are abundant in nature and can be derived from a variety of sources, including plants, animals, and microorganisms. Proteins and nucleic acids found in the human body, as well as cellulose, natural rubber, silk, and wool, are examples of natural polymers. Pharmaceuticals, tissue regeneration scaffolds, drug delivery, and imaging agents have all exploited modified natural polymeric materials. Alginate is a natural polymer that is common in brown seaweed (Phaeophyceae). It is an edible hetero-polysaccharide. (Draget and Taylor, 2011; ter Horst et al., 2019). According to researchers, they are indispensable in daily life, owing to their numerous applications. Because of its biodegradability and biocompatibility, alginate has a lot of potential in biomedical applications (Mollah et al., 2021).

Alginates are polymers of -(1-4)-Dmannuronic (M-blocks) and -L-guluronic acid (G-blocks) that are linear anionic polysaccharide polymers (Lee and Mooney, 2012). Alginate is usually extracted from brown algae by treating it with sodium hydroxide (NaOH) and then filtering it to speed up the synthesis of alginate. Through conversion and purification, water-soluble sodium alginate is created (Peteiro, 2017). As deterioration produces low molecular weight chemicals from natural polymers, radiation can break down alginates into smaller units of molecules (M-blocks and G-blocks) in some situations. Because of the number of GM blocks and their interchain interactions, alginate has gelling characteristics.

Edward Stanford, a researcher, was the first to employ alginates in 1883, and commercial development began in 1927. Alginates are now produced at a global rate of around 40,000 tonnes per year. In the food, pharmaceutical, cosmetic, and dentistry industries, alginates are widely utilized (Ertesvag and Valla, 1998). Perhaps the medical and pharmaceutical sectors have been increasingly prominent in biopolymers, particularly alginates, in recent years. This chapter gives a detailed account of the properties and applications of alginate for different applications.

Properties of Alginate

Alginates are insoluble substances that are washed, crushed, dried, powdered, and treated with a basic solution, such as NaOH or KOH, to produce a water-soluble sodium/potassium salt of alginic acid. Alginates are commonly found on the market as sodium alginate or alginic acid sodium salt (Soares et al., 2004; Lee et al., 2012). Alginates are a good biopolymeric biodegradable product in biomedical applications because of their acidic nature. Commercially accessible alginates are made from brown algae derived from seaweeds such as Laminaria hyperborea, Laminaria digitata, Laminaria japonica, Ascophyllum nodosum, and Macrocystis pyrifera (Figure 1) (Smidsrod, 2010). Alginates form gels quickly due to the presence of Guluronic acid (G) monomer in alginates, especially in the presence of Ca^{2+} ions, due to their high acid content. This gelling ability allows alginate to be used for a variety of purposes, including encapsulation of various fragments or even cells within the alginate matrix with few adverse effects (Klock et al., 1997). Alginates' carboxylic groups are particularly effective, which is why they have so many applications and may be changed as needed (Szabó et al., 2020). Bacterial alginate can be produced as an exopolysaccharide by bacteria such as Pseudomonas and Azotobacter. These bacterial alginate producers may be able to produce alginates with specific monomer formulations, as well as 'tailor-made' bacterial alginates through genetic and protein engineering (Lain et al., 2010).

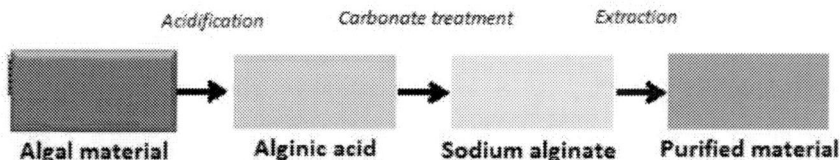

Figure 1. Procedure of extraction of purified sodium alginate from algal material.

The physicochemical heterogeneity of alginate has an impact on their quality. The same component also determines the potential relevance. Alginate comes in a variety of shapes, depending on the molecular weight (MW), arrangement, and distribution pattern of the two blocks (M & G). Physicochemical variables such as viscosity, sol/gel transition, and water uptake capabilities are some of the other parameters. Calculating the mean of all molecules present in the alginate sample reveals the molecular weight

(MW). The difference is between 33,000 and 400,000 g/mol. The M residues and block lengths vary depending on the source from which alginates are taken. In general, when the alginate G block matter or MW increases, the alginate gels become stronger and more fragile. Water and organic solvents do not dissolve amino acids. Monovalent alginate salts and alginate esters, on the other hand, are water-soluble and produce stable, viscous solutions (Remminghorst and Rehm, 2006; Cable, 2009). An aqueous solution (1 percent weight by volume of Na-Alg at 20°) has a dynamic viscosity of 20–400 mPa.s. The pH of the solvent, the strength of the ions, and the amount of gelling ions all limit the solubility of these compounds. A reduction in pH below pKa 3.38–3.65 may cause polymer precipitation (Kuo et al., 2001). In contrast to homogeneous structured molecules (Poly G and Poly M), which precipitate at lower pH, alginate with a more heterogeneous structure are water-soluble at lower pH (Draget et al., 1994; Braultet al., 2003). Depending on their concentration, solvent pH, the quantity of divalent ions present, and temperature, alginate compounds can produce viscous solutions. Due to their unique sol to gel transition capability, algae can be sculpted into a variety of semisolid/solid shapes under mild conditions. Alginate are used in the health and medicinal industries as viscosity enhancers, thickeners, and stabilizers for suspensions and emulsions. The eggbox model (Figure 2) (Shimokawa et al., 1996; Draget et al., 1994) can be used to bring about the sol to gel transition in Alginate by using divalent ions to produce cross-linking of polymer chains. Gelation can also be achieved by lowering the pH below the pKa of alginate monomers. Lactones, such as d-glucono-δ-lactone, can help with this. Calcium chloride, which is a source of calcium divalent ions, can enable alginate to gel quickly and unconstrainedly. The pace of gelation is a key aspect of controlling the gelation process. When gelation is delayed, homogenous gel structures are formed, which are mechanically robust. Several methods are used to reduce the rate of gelation, such as using phosphate buffers (e.g., sodium hexametaphosphate). The phosphate buffers battle divalent calcium ions during the reaction of alginate –COOH groups, preventing the gelation process (Fu et al., 2011). Other methods include using less water-soluble calcium sulfate and calcium carbonate, which lessen the gelation process. Another factor that affects the rate of gelation is temperature. Ca^{2+} becomes less reactive at low temperatures (Crow et al., 2006). In recent years, a freeze-thaw approach has been created that works well for regulated gelation (Augst et al., 2006). Gelling qualities are directly related to the alginate structure and proportions of M-, G-, and MG blocks. Furthermore, because of the increased number of repeating G block units, alginate gels are regarded as structurally stable, brittle, and stiffer. Alginate, on the other hand, produces flexible, somewhat elastic gels over time due to its high concentration of M-blocks. However, MG blocks in alginate gel determine its shrinkage and flexibility. Due to the high water absorption, alginate with a significant M-block composition swaps ions more easily than alginate with a maximal proportion of G-block content (Jogensen et al., 2007; Zhao et al., 2016).

The gelation process occurs under physical conditions, according to several investigations. Nonwoven calcium alginate dressings, for example, are used to treat oozing wounds or surgical injuries with infections by exchanging ions with the wound fluid (Doyle et al., 1996; Niekraszewicz et al., 2009; Wahl et al., 2015). The extremely absorbent soluble gel that is generated generates a wet environment and thus helps in healing by encouraging the new epidermis formation. Alginate are effective as structural supporting biomaterials for tissue healing due to their mechanical toughness and viscous nature. Alginate can undergo an in-site sol/gel transition, making them useful in several applications such as injectable vehicles in drug delivery and tissue engineering (Thomas, 2000). They can also be used as taste-making agents due to their gelling capabilities (Almeida et al., 2014; Rajesh et al., 2014). Spray-dried microspheres in combination with sodium alginate have been proven in studies to mask the bitter taste of ranitidine hydrochloride by forming a physical gel barrier (Jelvehgari et al., 2014). The swelling properties of alginate three-dimensional hydrogel structure are due to hydrophilic functional groups (Ciosek et al., 2015). The ability of alginate to produce gels and aid in the hydration process allows for a prolonged release of the active component at the admin site. As a result, these compounds are valuable in drug delivery systems for limited substance release (Tonnesen and Karlsen, 2002; Lee and Mooney, 2012). The alginate compound operate as advantageous compounds for trapping cells in tissue engineering and reconstruction because of the non-severe conditions in gelation (Ching et al., 2008; Sun and Tan, 2013; Garcia-Gareta et al., 2013). The materials that are immobilized are protected from physical stress by alginate barriers. This is done to ensure the long-term sustainability of the culture. Specialized alginate are being developed to treat a variety of disorders such as Parkinson's disease, diabetes, and cancer (Zhang et al., 2011; Calafiore et al., 2014). Alginate have good mucoadhesive qualities due to free –COOH groups that allow the polymer to join with mucin via hydrogen, causing electrostatic bonding. The solubility of alginate and the mucoadhesive nature of illnesses are both affected by the pH of the environment. Mucosal tissue can only bind with –COOH ions. Furthermore, soluble alginate facilitates the penetration of solvent into a polymer matrix, resulting in the formation of cohesive and viscous gel structures. Stronger mucoadhesive bonds result from the structure. Due to the reduction of alginate functional groups that bond with the mucin, excessive hydration of alginate matrix may cause a weakening of muco adhesiveness (Mythri et al., 2011; Hegarty et al., 2014; Haugstad et al., 2015). The various properties of alginate are presented in Figure 3.

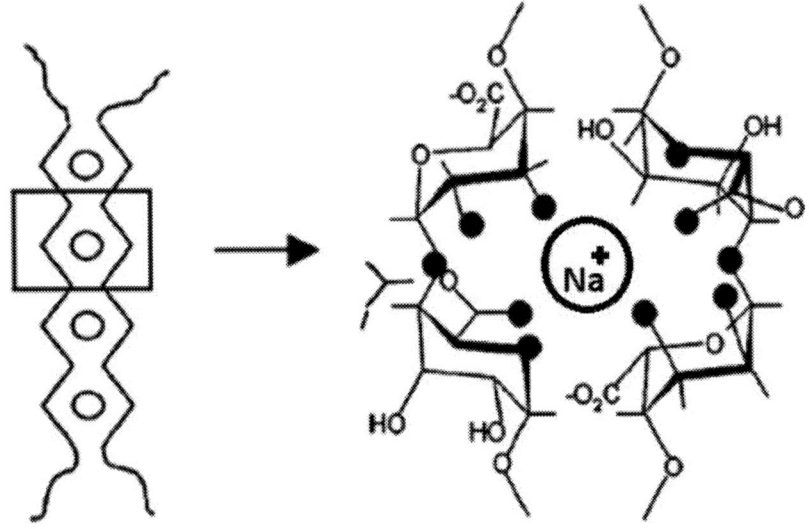

Figure 2. Diagrammatic representation of egg-box model.

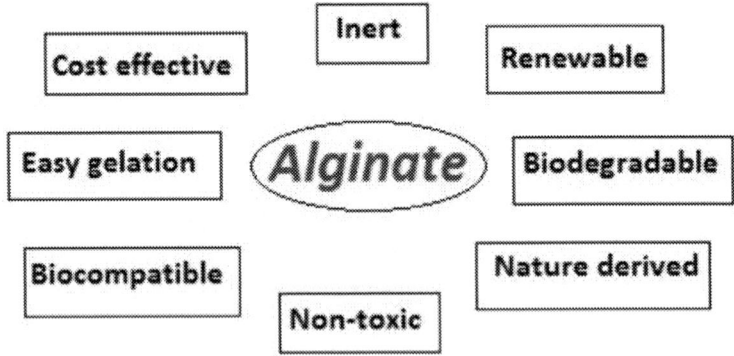

Figure 3. Essential properties of alginate.

Derivatives of Alginate

Alginates are used or synthesized for a variety of biomedical applications by incorporating different hydrophilic moieties into the alginate matrix, such as alkyl groups or hydrophilic polymers. Esterification bonds long-chain alkyl groups such as dodecyl or octadecyl to the matrix of alginates. Rheology, gelling, and crosslinking properties, for example, are extremely beneficial in bone regeneration and cartilage repair (Pelletier et al., 2001). Alginate derived from Poly (butyl methacrylate) is used to create sustained or controlled drug delivery vehicles (Yao et al., 2010). Alginates are also being studied for their cell-adhesive peptide derivatives, which are created by adding peptides as side chains and coupling them through the carboxylic groups of the sugar residues (Lehenkari and Horton, 1999; Koo et al., 2002).

Bioadhesion

Bioadhesion is the binding or contact between two surfaces, one of which is a biological substrate (Peppas and Buri, 1985). The mucosal layer is employed in mucoadhesion, for example. A mucoadhesive anionic polymeric layer is formed by the carboxyl group in alginates. Polyanion polymers are more effective bioadhesives than polycation or non-ionic polymers (Serp et al., 2000). Alginate has a higher mucoadhesive strength than polymers such as polystyrene, chitosan, carboxymethyl cellulose, and polyethylene glycol (lactic acid). Alginate's bioadhesive qualities make it a good mucosal drug delivery vehicle for the GI tract and nasopharynx because it prolongs drug residence time at the site of action, making it more efficacious (Gombotz and Wee, 1998; Gaserod et al., 1998; Bernkop-Schnurch et al., 2001).

Biocompatibility

Alginate's biocompatibility has been thoroughly tested *in vitro* and *in vivo* at various levels of purity. Alginate containing high M monomers has been reported to be more immunogenic and 10 times more effective in promoting cytokine synthesis than G monomer in alginates (Otterlei et al., 1991), but others have reported very little or no immune response across alginate implants (Zimmermann et al., 1992). Heavy metals, endotoxins, proteins, and polyphenolic compounds in the alginates may be causing the variable reaction at the injection or implantation sites. However, few serious inflammatory outcomes have been reported in commercially available or certified alginates obtained from branded companies (Lee et al., 2009). The quantity and quality of alginic acid determine its biocompatibility and strength. Furthermore, the effects of the amount of G monomer on alginate biocompatibility are still being studied. Experts differ on the G content of extremely purified alginate rich in G monomer residues, with some emphasizing high purity while ignoring the effects of chemical composition (Stabler et al., 2001). Animals such as rats are injected with calcium alginate in their kidneys for biocompatibility studies, and the results are very promising (Becker et al., 2001). Mammals have also been given alginates and found that they are unable to digest them because they lack the enzyme 'Alginase,' which breaks the polymer chains. Ionically cross-linked alginate gels dissolve by releasing monovalent cations from the divalent crosslinked gel into the surrounding fluids. Even though alginates disintegrate in the body, they are not evacuated because their molecular weight is greater than renal clearance (Shamkhani et al., 1995). Alginates derived from Undaria pinnatifida, a brown seaweed that has become invasive along the Argentine coast, are toxic, although purification using commercial procedures enhances biocompatibility and removes cytotoxicity in an alginate matrix for bone tissue engineering (Torres et al., 2019).

pH Effect

The solubility of alginates is affected by variables such as pH and ionic strength. Alginates have very low solubility in lower pH values due to carboxylic group deprotonation (–COO-). The viscosity of alginates is unaffected above pH > 5, whereas in solutions with pH 5, the COO- group in alginates is protonated to –COOH, and the electrostatic repulsion between chains decreases, allowing them to move closer together to form hydrogen bonds, resulting in a decrease in viscosity (Liu et al., 2002). De-polymerization reduces the viscosity of alginates with a pH greater than 11 (Mahmoodi et al., 2013). The concentration of an ionic solution affects crosslinking, which increases alginates' viscosity and molecular weight (Martisen et al., 1989; da Silva et al., 2017). Furthermore, crosslinking is dependent on the presence of monomers G and M groups in the alginate matrix.

Gelling Properties

Aqueous alginate solutions' ability to form gels when treated with divalent ions (Ca^{2+}, Sr^{2+}, and Ba^{2+}) or trivalent ions (Fe^{3+} and Al^{3+}) has been extensively investigated for the fabrication of carriers for sustained or controlled delivery of therapeutic agents which could be due to intramolecular bonding and ionic interactions between the carboxylic acid groups on the polymer matrix and the cations present. Calcium or other divalent or trivalent ions will interact with the G monomer in the alginate structure to crosslink with another molecule, and the structure is identical to the egg box model (Figure 2) (Grant et al., 1973). Complexing generating chemicals like EDTA-sodium citrate (Smeds and Grinstaff, 2001) or monovalent cations, complex anions (phosphate and citrate) with a high affinity for Ca2+ions, can easily rupture calcium alginate gels. The presence of non-gelling ions in high concentrations (Na+ and Mg2+) also contributes to the instability. It has been reported that the type of crosslinking and temperature has a large influence on the strength and uniformity of gelling (Kuo and Maa, 2001). It has also been reported that the strength of an alginate film crosslinked with Al3+ ions is very low when compared to other divalent ions (Ca2+ and Ba2+) crosslinking because crosslinking with Al3+ ions occurs in two different planes of the alginate structure while also making the alginate framework more compact (Al-Musa et al., 1999). The Al3+ ion's modest size (0.58 A) allows it to diffuse into the film matrix without crosslinking on the surface, resulting in poor crosslinking (Reddy et al., 2016). Alginates are being studied for covalent bonding to enhance the physical properties of gels for a range of applications, including tissue engineering. When alginates with covalent bonding are exposed to a modest increase in temperature, the chain degrades due to the breaking of crosslinks, resulting in stress relief due to water migration. It has been discovered that covalently linked alginates are poisonous (Zhao et al., 2010). The Ca2+ ions produced from the gel can also help with hemostasis, as the gel acts as a matrix for platelet and erythrocyte aggregation (Suzuki et al., 1998). Covalent cross-linking of alginate with Poly (ethylene glycol)-diamines of various molecular weights were first researched to generate gels with a wide range of mechanical properties. The elastic modulus of polyethylene glycol (PEG) gels improved with crosslinking density or weight fraction but then declined as the molecular weight between cross-links grew, compared to the original PEG (Eiselt et al., 1998). It was later discovered that the mechanical qualities, as well as the swelling of alginate strength, are closely managed using various types of cross-linking molecules and regulating cross-linking densities. The chemistry of the cross-linking molecules, as one might expect, has a significant impact on hydrogel swelling. Multi-functional cross-linking entities have a wider range of degradation efficiency and mechanical strength performance than bi-functional cross-linking molecules when it comes to molding hydrogels. *In vitro* studies were conducted on the physical characteristics and degradation behavior of Poly-aldehyde guluronate (PAG) gels made using either Poly-acrylamide-co-hydrazide (PAH) or Adipic acid dihydrazide (AAD) as a cross-linking agent. In comparison to PAG/AAD gels, PAG/PAH gels demonstrated superior mechanical rigidity and very little degradation (Lee et al., 2004). Photo crosslinking is used to crosslink alginates structures in the presence of suitable initiators or under mild conditions such as direct or indirect sunlight. This technique, for example, significantly improved the mechanical properties of Polyallylamine and alginate (Lu et al., 2000).

Immunogenicity

In the presence of pharmaceutical dosage requirements for successful application in drug carriers, controlled drug delivery is the latest trend; alginates play an important role due to their biocompatibility and immunogenicity (Gombotz and Wee, 1998). Alginate immunogenicity is caused by two factors: its chemical composition and the presence of mitogenic pollutants in alginates. Alginate is thought to have mild cytotoxic effects and reduce hemolysis when it comes into contact with blood (Becker et al., 2001).

Molecular Weight

Alginates obtained from various seabed locations have molecular weights ranging from 50,000 to 5 lakhs (Lakshmi et al., 2007). Because carboxylate groups in Guluronic acid present in the alginate structure protonate and form hydrogen bonds, the viscosity of the alginate solution is pH-responsive and increases viscosity with decreasing pH (Rinaudo, 1992). Alginates may have different molecular weights depending on whether pre-gel solution viscosity or post-gelling strength distribution must be monitored separately. A mixture of high and low molecular weight alginate polymers is used to adjust the viscosity of the solution (Kong et al., 2002).

Sterilization

The viscosity of alginates is said to decrease with autoclave sterilization because the heating randomly breaks alginate chains. The amount of loss is determined by the presence of other types of stuff in the solution. -radiation and ethylene oxide have also been used to sterilize alginate solutions (Vandenbossche and Remon, 1993).

Solubility

Because alginates contain a terminal carboxylic ion (–COO-), the divalent or trivalent cations bond to this and produce an insoluble product. As a result, the alternative is for them to absorb up to 200–300 times their weight in water, swelling to a paste-like hydrogel. Alginates containing monovalent cations (Na^+, K^+, and NH_4^+) are soluble in both hot and cold water. Because of their different molecular weights, alginates have a wide range of solubilities. For example, alginates derived from Ascophyllum have aqueous solubility in the range of 22–30% weight percent, whereas those of two the Laminaria groups are 17–33% weight percent and 25–44% weight percent, respectively (Das and Senapati, 2008; da Silva et al., 2017).

Toxicity

Numerous investigations have found that alginates, particularly cross-linked sodium/calcium alginates, are non-toxic to cells and do not cause eye or skin irritation. Because of their nontoxicity, they've found uses in the pharmaceutical, cosmetics, and food industries (Dusel et al., 1986).

Applications of Alginate

Alginates are abundant in the oceans, and they have several applications in the food business, pharmaceuticals, cosmetics, textile industry, welding, and animal feeds, among others, due to their different qualities such as biodegradation, biodegradable, non-toxic, and so on. They belong to the polyanionic copolymer family, which is derived from marine kelp, particularly brown sea algae.

Table 1. Different types of commercially available alginates and their industrial use

S.No.	Types of alginates	Industrial applications	Reference
1.	Alginic acid	Emulsifier, formulation aid, stabilizer, thickener, Tablet binder and disintegrant, sustained release and release-modifying agent, taste masking agent, suspending and viscosity increasing agent	Repka and Singh, 2009
2.	Ammonium alginate	Stabilizer, thickener, humectant, Color diluent, emulsifier, film former, humectant	Shahand and Thassu 2009
3.	Calcium alginate	Stabilizer, thickener, Tablet disintegrant	Cable, 2009
4.	Propylene glycol alginate	Emulsifier, flavoring adjuvant, formulation aid, stabilizer, surfactant, thickener, Stabilizer, emulsifier, suspending and viscosity increasing agent	Nause et al., 2009
5.	Sodium alginate	Texturizer, stabilizer, thickener, formulation aid, firming agent, flavour adjuvant, emulsifier, surface active agent, Suspending and viscosity increasing agent, tablet and capsule disintegrant, tablet binder, sustained release agent, and diluent in capsule formulation,	Cable C. G., 2009

Stanford patented impure alginate, which was studied extensively due to its unique properties as an anionic polysaccharide. Alginate is utilized as a thickener, stabilizer, and in various items such as jellies, beverages (chocolate milk), and sweets (ice cream). For the production of ethanol, yeast cells are encapsulated with alginate. Alginate can be used as a thickening agent for color pigments in fabric printing, as an adhesive agent, and as a filler in the paper industry. It can also be used in paintings as a stabilizer and suspending agent. This improves painting flow while also increasing the surface coverage. Alginate is employed in ceramic shaping and water treatment as biocatalysts (Wang et al., 2002). They could potentially be used to sequester metal (Aderhold et al., 1996). Alg pellets trapped in the fungus can react with harmful metal ions in a specific way. These are employed in the sewage and water treatment sectors (Arlca et al., 2001). Alginates are also used in the healthcare and pharmaceutical industries for transplantation and cell culture

(Lim et al., 1980). Hybridoma cells are used in monoclonal antibody synthesis and drug immobilization for controlled release (Molly et al., 2004). Alginate in tablet development can also improve the bioadhesion property of buccal sticky tablets (Wang et al., 2010). Diabetes and hypercholesterolemia can be treated with sodium alginate with a low molecular weight (MW) of 10–100 kDa. This is accomplished by encapsulating blood glucose and cholesterol, then gelling amino acids in the stomach (Choi et al., 2000). Alginate functional responsibilities in a range of applications are depicted in Table 1.

Pharmaceutical Application

In today's medical world, many medicines and therapies that are given to patients have several negative effects. As a result, there is a high need for drug loading carriers that increase drug resident time *in vitro* or *in vivo*, particularly in the gastrointestinal system and on the surface of the human body. These carriers must be safe and non-toxic. Controlled drug delivery, sustained drug delivery, and targeted drug delivery are all considered in relation to alginates and alginate derivatives.

In general, the structure of alginates gels demonstrates that it has a porous size (5 nm), which aids in filling this gap with small molecular weight drugs via physical or chemical bonding. The drug release is controlled when a drug-loaded/embedded drug comes into contact with an aqueous medium. Furthermore, because the drug-loaded carriers are water-soluble and may degrade in an aqueous medium, crosslinking the alginates with bivalent or trivalent cation will improve the gels or films' stability. These properties aid in the study of drug release kinetics. In-vitro controlled release of Valganciclovir hydrochloride as an anti-HIV drug is being investigated using the sodium salt of alginic acid (SA) and polyethylene oxide blends (Mallikarjuna et al., 2013). Metformin hydrochloride medication administration was studied using floating microbeads composed of SA and modified Chinese yam starch (Okunlola et al., 2010). For controlled release, SA and Chitosan blend with various wt% Montmorillonite (Cloisite 30B) solutions were investigated (Nayak et al., 2011). In-vitro release of an anti-cancer medicine such as Paclitaxel was examined in a changing pH medium, duration, and drug concentration. Ranitidine hydrochloric acid medication *in vitro* release was examined in simulated intestinal fluid (pH 7.5) and simulated gastric fluid (pH 1.2) using SA and Xanthum gum mixes crosslinked with zinc acetate loading (Zeng et al., 2004). Using composite microparticles constructed of Chitosan, SA, and Pectin crosslinked by tripolyphosphate, oral administration of protein drugs for Bovine serum albumin (BSA) is investigated (Yu et al., 2009). The regulated release of Diclofenac sodium is investigated using a pH-responsive Tamarind seed polysaccharide and alginate combination. Swelling and degradation tests in various pH mediums were also conducted (Nayak and Pal, 2011). For microspheres produced by SA and Methylcellulose utilizing Glutaraldehyde as a crosslinking agent, an anti-inflammatory medication such as Nifedipine is studied (Babu et al., 2007). On exposure to UV radiation, the Acrylamide and Poly (vinyl alcohol) beads containing SA are grafted and further crosslinked using glutaraldehyde. The regulated release of Diclofenac sodium medication is studied using crosslinked beads (Sanl et al., 2007). SA and Xanthum gum are used to make transdermal films. Skin penetration of films containing Ketoprofen was investigated *in vitro* (Rajesh et al., 2010). SA in combination with Sodium Carboxymethylcellulose, Carbopol-934, Polyvinyl pyrrolidine, Ethylcellulose as a supporting membrane, and Glycerol as a plasticizer yields excellent results for drug control (Chaudhary et al., 2010). In simulated intestinal media, hypertension medications like Felodipine were tested for control release from alginate microspheres in combination with a mixture of Hydroxypropyl methylcellulose, Eudragit RS 30D, and Chitosan (Lee et al., 2003). In-vitro release tests were carried out on a bioadhesive ocular insert containing Ciprofloxacin hydrochloride employing SA as a gel, Chitosan as a bioadhesive agent, and Glycerin as a plasticizing agent (Shinde et al., 2014).

Application in Protein Delivery

Protein medications are in high demand, and breakthroughs in recombinant DNA technology have made a wide range of protein drugs available. Alginates are an excellent choice for protein drug delivery because they can load into the alginate matrix in a variety of formulations while avoiding denaturation and degradation. The regulated release of protein from alginate gels is investigated using several approaches. Because of their bi-polymeric structural arrangements, alginates are recognized for having small pores due to the presence of G and M blocks. The sustained and localized release of vascular endothelial growth factors from alginate hydrogels has been demonstrated (Lee et al., 2003; Silva et al., 2010). Alginate microspheres are effective at loading lysozyme and chymotrypsin for long-term release (Wells et al., 2007). Amino group-terminated proteins, for example, are delivered orally (Gao et al., 2009). Poly((2-dimethylamino) ethyl methacrylate was mixed with alginate gel beads. In the fabrication of tetra-functional acetal-linked polymer networks with changeable pore widths, alginates are investigated for stimuli-responsive gels (Yu et al., 2009). Alginate, chitosan, and pectin composite mixtures are employed as a model protein for bovine serum albumin. Hemoglobin was used as a model protein in poly (L-histidine)-chitosan/alginate microcapsules in another study (Chen et al., 2012; Sosnik, 2014).

Alginate in Cosmetics

For decades, scientists have studied alginates to develop high-quality cosmetic products that deliver all of the skin's benefits. Alginates are marine plants that absorb UV radiation, repair sun damage, moisturize the epidermis, smooth the skin, and ensure the renewal of tiny cells. Alginates are utilized in many cosmetics because of their ability to thicken and keep moisture. Alginate aids in the retention of lipstick color on the lips, as well as face creams and body lotion moisturizers, by producing a gel network. Alginate, a natural thickening, is added to sunflower wax to generate stable all-around lotions. Polysorbate-20 is the greatest commercial product, and it's a silky lotion that combines quick cold emission products with an emulsifier. Alginates are a type of natural polysaccharide with high viscosity and a high water absorption capacity. Perhaps the viscosity of alginate can be improved to get maximum viscosity. Anti-aging masks and face masks containing alginates are being researched to help slow down the aging process, smooth out wrinkles and lift the skin. Alginates are also investigated in Dentures, a removable set of replacement teeth and gum tissue that can offer you the entire support and attractive appearance you require as you age. Alginates application indentures are a removable set of restorative dentistry and gum tissues that can provide us with complete support and a lovely appearance even as we get older (Reddy et al., 2021; Raghav et al., 2021).

Application as Wound Dressings

Any damage, burns, tears, muscle pains, or cuts that occur on the human body may take longer to heal, and using antiseptic, antimicrobial, anti-inflammatory, antipruritic, pain-relieving gels, or anti-mycotic with fungal action may produce skin irritation or side effects. As a result, alginates are widely employed in wound healing to load relevant medications in alginate gels, which enhances the drug's retention period, allowing the drug to be released in small doses at the specified spot. Alginates also have hemostatic qualities; therefore, they can be used to heal bleeding wounds. Several studies have been published in which mice, pigs, and rabbits were used as test subjects. Alginates are used to create hydrogels, films, wafers, foams, nanofibers, and therapeutic wound dressing compositions. Alginate wound dressings absorb wound fluid quickly, maintain a physiologically moist environment, and protect the wound site from bacterial infections. Alginates are combined with other polymers or composites to improve film strength because they have low mechanical strength. M-block stimulates cytokine production and the amount of M-block present in alginate determines the immunogenic effect (Szekalska et al., 2016).

Application as Animal Feeds

Sodium and potassium alginate salts are intended for use as emulsifiers, stabilizers, thickeners, gelling agents, and binders in industry. There is no authoritative recommendation for using sodium alginate in feeding kinds of stuff for dogs, other non-food-producing animals, or fish. Potassium alginate, on the other hand, is utilized in cat and dog food at a concentration of 40 g/kg feed (Bastos et al., 2017). Alginates in fish feed have no negative impact on the consumer. Alginates are known to irritate the eyes slightly but not the skin. These components are not harmful to the aquatic environment when used in fish feed. To prevent the calcium content from reacting with the alginate, a gel-type animal feed mixture is made by mixing feed ingredients, water, alginate, and a water-insoluble calcium component. After the feed mixture is created, the calcium component is solubilized or the sequestrate impacting the reactivity between the alginate and the calcium component is removed, yielding a gel feed with a gel matrix holding the feed nutrient elements. The gel meal can subsequently be fed to the livestock (Lanter et al., 2012).

Application in Textile Industry

Patterns are applied to printed fabrics, shawls, towels, and other objects using color paste substrates made from textile-grade alginates. Other substrates for textile printing are cleaner and easier to degrade than alginates. The use of alginates in printing cotton, jute, and rayon provides for better wastewater disposal. Thickeners like sodium alginates are used in textile printing to thicken the dye paste. The pastes can be applied to the textile using screen or roller printing equipment. When reactive dyes were discovered, alginates became popular thickeners. These compounds react chemically with the cellulose in the textile. Many common thickeners, such as starch, react with reactive dyes, resulting in reduced color yields and often difficult-to-clean by-products. Alginates are the most effective thickeners for reactive dyes since they are non-reactive with the dyes and wash away rapidly. Alginates of medium to high viscosity were utilized in previous screen printing. However, with modern high-speed roller printers, even low viscosity alginates produce very appealing printing (Reddy et al., 2021).

Application in Welding Rods

Welding is a method used to construct metal structures of various types. Process coating is used as flux and to monitor conditions near the weld, such as temperature, oxygen, and hydrogen, during welding. To give some plasticity for coating extrusion into the rod and to attach the dried coating to the rod, sodium silicate (water glass) is added with the dry coating materials in this case. On the other hand, the silicate neither binds nor offers sufficient lubrication to allow for successful and smooth extrusion. To keep the moist mass together before extrusion and to retain the coating on the rod in shape throughout drying and baking, a lubricant and a binder are required. Alginates are employed to meet these requirements (Reddy et al., 2021).

Application in Food Industry

Because of their particular qualities in food applications, the FAO/WHO considers alginates to be the safest food additives. Because of its distinctive features such as thickening, gelling, emulsification, stability, and textural functioning, alginates have been utilized in a variety of culinary items for decades, including ice cream toppings, fruit jams, jelly, milk products, food packaging, instant noodles, beer, and so on (Shilpa et al., 2003). Alginate based modified hydrogels are presented in Table 2.

Table 2. Modified alginate based hydrogels and their applications

S.No.	Alginate based hydrogels	Uses	Reference
1.	Alginate-based nanocellulose	Wound-healing biotechnology	Siqueira et al., 2019
2.	Akermanite, alginate	Wound-healing, and bio-engineering	Yan et al., 2017
3.	Bioglass/agarose, alginate	Chronic Wound-healing	Zeng et al., 2015
4.	Carboxymethylcellulose (CMC), alginate, gatifloxacin	Antibacterial	Kesavan, 2010
5.	CMC incorporated with chitosan, alginate	Chronic wounds	Lv et al., 2019
6.	Polyacrylamide, alginate, cations (Cu^{2+}, Zn^{2+}, Sr^{2+}, Ca^{2+})	Wound-healing	Zhou et al., 2018
7.	Chitosan, alginate, alpha-tocopherol	Wound-healing	Ehterami et al., 2019
8.	CMC hydrogel, chitosan, cellulose nanocrystal	Burn wound-healing	Huang et al., 2018

Conclusion and Future Perspective

The interest in the creation of natural and biosynthetic materials for medical devices has grown significantly. Another, more in-depth experiment should be started right away to uncover possible medical applications. Alginates have a lot of potential as biomaterials for human healthcare and mainstream biotechnology applications such as tissue planting, drug delivery, wound healing, stem cell culture, and gene engineering. When compared to other commercially available conventional wound dressing materials, alginate-based biomaterials can be removed with less discomfort. Furthermore, because alginate-based sustained-release antibiotics have a high absorption capacity, they are ideal for treating deep burns. Biocompatibility, mild gelation, and moderate modification of alginate gels are the most effective physicochemical, mechanical, and thermal features that need to be improved for these applications to develop alginate derivatives with ultramodern characteristics. Finally, the world is changing; progressing through newly invented technology and high-tech facilities, but environmental and other aspects of sustainability are still limited in these sectors. Thus, the modification and characterization of alginate polymers and their biofilms in biomedical applications for diverse objectives, as well as limiting environmental consequences to allow for sustainable societies, could play a critical role in their development for the medical sector. The study's findings have the potential to expand the scope of healthcare applications in the medical field.

Conflict of Interest

The authors declare that they have no competing interests.

Acknowledgments

We sincerely thank Department of Biotechnology (DBT), India (Scientific research project BT/ PR 15650/ AAQ/3/815/2016) for providing financial support.

References

Aderhold, D., Williams, C. J., Edyvean, R. G. J., The removal of heavy-metal ions by seaweeds and their derivatives. *Bioresour. Technol.*, 58, 1, 1996.

Alejandro Sosnik, Alginate Particles as Platform for Drug Delivery by the Oral Route: State-of-the-Art International Scholarly Research Notices. 2014, Article ID 926157. http://dx.doi.org/10.1155/2014/926157

Almeida, H., Amaral, M. H., Lobao, P., Lobo, J. M. S., *In situ* gelling systems: A strategy to improve the bioavailability of ophthalmic pharmaceutical formulations. *Drug Discovery Today*, 19, 400, 2014.

Al-Musa S., Fara, D. A. and Badwan A. A., Evaluation of parameters involved in preparation and release of drug loaded in crosslinked matrices of alginate J. Controlled Release, 57, 223 (1999). doi: 10.1016/s0168-3659(98)00096-0.

Anu Shilpa, S. S. Agrawal; Alok R. Ray, *Polymer Reviews*. 2003;43(2):187 -221. doi: 10.1081/MC- 120020160.

Arlca, M. Y., Kacar, Y., Genc, O., Entrapment of white-rot fungus *Trametes versicolor* in Ca-alginate beads: Preparation and biosorption kinetic analysis for cadmium removal from an aqueous solution. *Bioresour. Technol.*, 80, 121, 2001.

Augst, A. D., Kong, Mooney, D. J., (2006) Alginate hydrogels as biomaterials. *Macromol. Biosci.*, 6, 623.

Becker T. A., Kipke D. R., Brandon T. Calcium alginate gel: A biocompatible and mechanically stable polymer for endovascular embolization. *J. Biomed. J Biomed Mater Res.* 2001;54(1):76-86. doi: 10.1002/1097-4636(200101)54:1<76::aid-jbm9>3.0.co;2-v.

Bernkop-Schnurch A., Kast C. E., Richter M. F. Improvement in the mucoadhesive properties of alginate by the covalent attachment of cysteine. *J. Control Release.* 2001;71(3):277-285. doi.org/10.1016/S0168-3659(01)00227-9.

Brault D., Heyraud A., Lognone V., Roussel M. (2003) Methods for obtaining oligo mannuronates and guluronates, products obtained and use thereof, Patent WO03099870, 2003.

Cable, C. G., Sodium alginate, in: *Handbook of Pharmaceutical Excipients*, 6th edition, p. 622, Pharmaceutical Press, London, UK, 2009.

Cable, C. G. "Calciumalginate," in *Handbook of Pharmaceutical Excipients*, pp. 83–85, Pharmaceutical Press, London, UK, 6th edition, 2009.

Cable, C. G. "Sodium alginate," in *Handbook of Pharmaceutical Excipients*, pp. 622–624, Pharmaceutical Press, London, UK, 6th edition, 2009.

Calafiore, R. and Basta, G., Clinical application of microencapsulated islets: Actual prospectives on progress and challenges. *Adv. Drug Deliv. Rev.*, 67, 84, 2014.

Chen A. Z., Chen M. Y., Wang S. B., Huang X. N., Liu Y. G., Chen Z. X. Poly(L-histidine)-chitosan/alginate complex microcapsule as a novel drug delivery agent. *Jol of App Polymer Science.* 2012; 124(5): 3728-3736. Doi.org/10.1002/app.35371

Ching, A. L., Liew, C. V., Chan, L. W., Heng, P. W. S., Modifying matrix microenvironmental pH to achieve sustained drug release from highly laminating alginate matrices. *Eur. J. Pharm. Sci.*, 33, 361, 2008.

Choi, H. G. and Kim, C. K., Development of omeprazole buccal adhesive tablets with stability enhancement in human saliva. *J. Control. Release*, 68, 397, 2000.

Ciosek, P., Wesoły, M., Zabadaj, M., Lisiecka, J., Sołłohub, K., Cal, K., & Wróblewski, W. Towards flow through/flow injection electronic tongue for the analysis of pharmaceuticals. *Sens. Actuators B: Chem.*, 207, 1087, 2015.

Crow, B. B. and Nelson, K. D., (2006) Release of bovine serum albumin from a hydrogel-cored biodegradable polymer fiber. *Biopolymers*, 81, 419.

Cui-Yun Yu, Bo-Cheng Yin, Wei Zhang, Si-Xue Cheng, Xian-Zheng Zhang, Ren-Xi Zhuo. Composite microparticle drug delivery systems based on chitosan, alginate and pectin with improved pH-sensitive drug release property. *Colloids and Surfaces B: Biointerfaces*, 2009;68 (2):245.doi.org/10.1016/j.colsurfb.2008.10.013.

da Silva T. L., Vidart J. M., da Silva M. G., Gimenes M. L., Vieira M. G. Alginate and Sericin: Environmental and Pharmaceutical Applications. In *Biological Activities and Application of Marine Polysaccharides;* InTech: Rijeka, 2017; pp 57-86. doi: 10.5772/65257.

Das M. K., Senapati P. C. Furosemide-loaded Alginate microspheres prepared by ionic cross-linking technique: Morphology and release characteristics. *Indian J. Pharm. Sci.* 2008, 70, 77-84. doi: 10.4103/0250-474X.40336.

Doyle, J. W., Roth, T. P., Smith, R. M., Li, Y. Q., Dunn, R. M., Effect of calcium alginate on cellular wound healing processes modeled *in vitro. J. Biomed. Mater. Res.*, 32, 4, 561, 1996.

Doyle, J. W., Roth, T. P., Smith, R. M., Li, Y. Q., Dunn, R. M., Effect of calcium alginate on cellular wound healing processes modeled *in vitro. J. Biomed. Mater. Res.*, 32, 4, 561, 1996.

Draget K. I., and Taylor C. (2011). Chemical, Physical and Biological Properties of Alginates and Their Biomedical Implications. *Food Hydrocolloids* 25, 251–256. doi:10.1016/j.foodhyd.2009.10.007.

Draget, K. I., Simensen, M. K., Onsoyen, E., Smidsrod, O., (1993) Gel strength of Ca-limited alginate gels made *in situ*. *Hydrobiologia*, 260–261, 563.

Draget, K. I., Skjak Brak, G., Smidsrod, O (1994) Alginic acid gels: The effect of alginate chemical composition and molecular weight. *Carbohydr. Polym.*, 25, 1, 31.

Dusel R., McGinity J., Harris M. R., Vadino W. A., Cooper J. Sodium alginate. In *Handbook of Pharmaceutical Excipients;* Published by American Pharmaceutical Society: USA and The Pharmaceutical Society of Great Britain: London. 1986; 257-258.

Ehterami, A., Salehi, M., Farzamfar, S., Samadian, H., Vaez, A., Ghorbani, S., Ai, J., & Sahrapeyma, H. (2019). Chitosan/alginate Hydrogels Containing Alpha-Tocopherol for Wound Healing in Rat Model. *J. Drug Deliv. Sci. Tech.* 51, 204–213. doi:10.1016/j.jddst.2019.02.032.

Eiselt P., Lee K. Y., Mooney D. J. Rigidity of two-component hydrogels prepared from alginate and poly(ethylene glycol)-diamines. *Macromolecules* 1999;32:5561-6. doi.org/10.1021/ma990514m.

Ertesvag H, Valla S (1998): Biosynthesis and applications of alginates. Polymer Degradation and Stability, 5, 985

Fu, S., Thacker, A., Thacker, A., Sperger, D. M., Boni, R. L., Buckner, I. S., Velankar, S., Munson, E. J., Block, L. H., (2011) Relevance of rheological properties of sodium alginate in solution to calcium alginate gel properties. *AAPS PharmSciTech*, 12, 453.

Gao C. M., Liu M. Z., Chen S. L., Jin S. P., Chen J. Preparation of oxidized sodium alginate-graftpoly((2- dimethylamino) ethyl methacrylate) gel beads and *in vitro* controlled release behavior of BSA. *Int J Pharm*. 2009; 371:16-24. doi: 10.1016/j.ijpharm.2008.12.013.

Garcıa-Gareta, E., Ravindran, N., Sharma, V., Samizadeh, S., Dye, J. F., A novel multiparameter *in vitro* model of three dimensional cell ingress into scaffolds for dermal reconstruction to predict *in vivo* outcome. *Biores. Open Access*, 2, 412, 2013.

Gaserod, O, Jolliffe I. G, Hampson F. C., Dettmar P. W., Skjak-Braek G. The enhancement of the bioadhesive properties of calcium alginate beads by coating with chitosan. *Int. J. Pharm*. 1998; 175 (2): 237-246. doi.org/10.1016/S0378-5173(98)00277-4.

Gombotz, W. R., Wee, S. F. Protein release from alginate matrices. *Adv. Drug Deliv. Rev.* 1998, 31 (1- 2), 267-285. doi: 10.1016/s0169-409x(97)00124-5.

Grant G. T., Morris E. R., Rees D. A., Smith P. J. C., Thom D. Biological interactions between polysaccharides and divalent cations: The egg-box model. *FEBS Letters*. 1973; 32 (1):195-198. doi.org/10.1016/0014-5793(73)80770-7.

Haugstad, K., Håti, A., Nordgård, C., Adl, P., Maurstad, G., Sletmoen, M., Draget, K., Dias, R., & Stokke, B. Direct determination of CS–mucin interactions using a single molecule strategy: Comparison to alginate–mucin interactions. *Polymers*, 7, 2, 161, 2015.

Hegarty, S. V., Keeffe, G. W. O., Sullivan, A. M., Neurotrophic factors: From neuro developmental regulators to novel therapies for Parkinson's disease. *Neural Regen. Res.*, 9, 1708, 2014.

Huang, W., Wang, Y., Huang, Z., Wang, X., Chen, L., Zhang, Y., & Zhang, L. (2018). On-Demand Dissolvable Self-Healing Hydrogel Based on Carboxymethyl Chitosan and Cellulose Nanocrystal for Deep Partial Thickness Burn Wound Healing. *ACS Appl. Mater. Inter.* 10 (48), 41076–41088. doi:10.1021/acsami.8b14526.

Jelvehgari, M., Barghi, L., Barghi, F., Preparation of chlorpheniraminemaleate-loaded alginate/CS particulate systemsby the ionic gelation method for taste masking. *Jundishapur J. Nat. Pharm. Prod.*, 9, 39, 2014.

Jorgensen, T. E., Sletmoen, M., Draget, K. I., Stokke, B. T., (2007) Influence of oligoguluronates on alginate gelation, kinetics, and polymer organization. *Biomacromolecules*, 8, 2388.

Kent Lanter, Brenda de Rodas, Bill L. Miller, Gary E. Fitzner, US Patent US 8092853 B2., Jan 2012. doi: 10.2903/j.efsa.2017.4945.

Kesavan, K. (2010). Sodium Alginate Based Mucoadhesive System for Gatifloxacin and its *In Vitro* Antibacterial Activity. *Sci. Pharm.* 78 (4), 941–957. doi:10.3797/scipharm.1004-24.

Klöck, G., Pfeffermann, A., Ryser, C., Gröhn, P., Kuttler, B., Hahn, H.-J., & Zimmermann, U.Biocompatibility of mannuronic acid-rich alginates. *Biomaterials.* 1997, 18:707-713. doi: 10.1016/s0142 9612(96)00204-9.

Kong H. J., Lee K. Y., Mooney D. J. Decoupling the dependence of rheological/mechanical properties of hydrogels from solids concentration. *Polymer.* 2002; 43:6239-6246. doi.org/10.1016/S0032- 3861(02)00559-1.

Koo L. Y., Irvine D. J., Mayes A. M., Lauffenburger D. A., Griffith L. G. Co-regulation of cell adhesion by nanoscale RGD organization and mechanical stimulus. *J Cell Sci.* 2002; 115:1423-1433.

Kuo, C. K. and Ma, P. X., Ionically crosslinked alginate hydrogels as scaffolds for tissue engineering: Part 1. Structure, gelation rate and mechanical properties. *Biomaterials*, 22, 6, 511, 2001.

Lain Hay D, Zahid Ur Rehman, Aamir Ghafoor, Bernd H. A. Rehm, Bacterial biosynthesis of alginates, *Journal of chemical tech and Biotech.* 2010;85(6): 752-759. doi.org/10.1002/jctb.2372.

Lakshmi S. Nair, Cato T. Laurencin. Biodegradable polymers as biomaterials. *Prog. Polym. Sci.* 32 (2007) 762-798. doi:10.1016/j.progpolymsci.2007.05.017.

Lee D. W., Hwang S. J., Park J. B., Park H. J. Preparation and release characteristics of polymer-coated and blended alginate microspheres. *Journal of Microencapsulation.* 2003; 20(2): 179-192. doi.org/10.3109/02652040309178060.

Lee J, Lee KY. Local and sustained vascular endothelial growth factor delivery for angiogenesis using an injectable system. *Pharm Res* 2009; 26:1739-1744. doi: 10.1007/s11095-009-9884-4.

Lee K. Y., and Mooney D. J. (2012). Alginate: Properties and Biomedical Applications. *Prog. Polym. Sci.* 37 (1), 106–126. doi:10.1016/ j.progpolymsci.2011.06.003.

Lee K. Y., Bouhadir K. H., Mooney D. J. Controlled degradation of hydrogels using multifunctional cross-linking molecules. *Biomaterials,* 2004;25:2461-6. doi.org/10.1016/j.biomaterials.2003.09.030.

Lee K. Y., Peters M. C., Mooney D. J. Comparison of vascular endothelial growth factor and basic fibroblast growth factor on angiogenesis in SCID mice. *J Control Release* 2003; 87:49-56. doi: 10.1016/s0168-3659(02)00349-8.

Lee, K. Y. and Mooney, D. J., Alginate: Properties and biomedical applications. *Prog. Polym. Sci.*, 37, 1, 106, 2012. Tonnesen, H. H. and Karlsen, J., Alginate in drug delivery systems. *Drug Dev. Ind. Pharm.*, 28, 621, 2002.

Lehenkari P. P., Horton M. A. Single integrin molecule adhesion forces in intact cells measured by atomic force microscopy. *Biochem Biophys Res Commun.* 1999; 259:645-650. https://doi.org/10.1006/bbrc.1999.0827.

Lim, F. and Sun, A. M., Microencapsulated islets as bioartificial endocrine pancreas. *Science*, 210, 908, 1980.

Liu, X. D., Yu, W. Y., Zhang, Y., Xue, W. M., Yu, W. T., Xiong, Y., Ma, X. J., Chen, Y., Yuan, Q. Characterization of structure and diffusion behaviour of Ca-Alginate beads prepared with external or internal calcium sources. *J. Microencapsul.* 2002, 19, 775-782. doi.org/10.1080/02652040210000227243.

Lu M. Z., Lan H. L., Wang F. F., Chang S. J., Wang Y. J. Cell encapsulation with alginate and phenoxycinnamylidene-acetylated poly(allylamine). *Biotechnol Bioeng* 2000;70:479- 83.doi.org/10.1002/1097-0290(20001205)70:5<479::AID BIT1> 3.0.CO;2-E.

Luca Szabó, Sandrine Gerber-Lemaire, Christine Wandrey, Strategies to functionalize the anionic biopolymer Na-Alginate without restricting its polyelectrolyte properties, *Polymers* (Basel). 2020; 12(4): 919. doi: 10.3390/polym12040919.

Lv, X., Liu, Y., Song, S., Tong, C., Shi, X., Zhao, Y., Zhang, J., & Hou, M. (2019). Influence of Chitosan Oligosaccharide on the Gelling and Wound Healing Properties of Injectable Hydrogels Based on Carboxymethyl Chitosan/alginate Polyelectrolyte Complexes. *Carbohydr. Polym.* 205, 312–321. doi:10.1016/ j.carbpol.2018.10.067.

Mahmoodi, NM. Magnetic ferrite nanoparticle–Alginate composite: Synthesis, characterization an binary system dye removal. *J. Taiwan Inst. Chem. Eng* 2013, 44,322-330. doi.org/10.1016/j.jtice.2012.11.014.

Mallikarjuna B, Madhusudana Rao K, Siraj S, Chandra Babu A, Chowdoji Rao K, Subha, MCS. Sodium alginate/poly (ethylene oxide) blend hydrogel membranes for controlled release of Valganciclovir hydrochloride. *Designed Monomers and Polymers.* 2013;16(2);151. doi.org/10.1080/15685551.2012.705503.

Maria de Lourdes Bastos, Georges Bories, Andrew Chesson, Pier Sandro Cocconcelli, Gerhard Flachowsky, Boris Kolar, Maryline Kouba, Marta L_opez-Alonso, Secundino L_opez Puente, Alberto Mantovani, Baltasar Mayo, Fernando Ramos, Maria Saarela, Roberto Edoardo Villa, Robert John Wallace, Pieter Wester, Anne-Katrine Lundebye, Carlo Nebbia, Derek Renshaw, Safety and efficacy of sodium and potassium alginate for pets, other nonfood producing animals and fish. *EFSA Journal* 2017; 15(7):4945.

Martinsen, A., Skjåk-Braek, G., Smidsrød, O. Alginate as immobilization material: I. correlation between chemical and physical properties of Alginate gel beads. *Biotechnol. Bioeng.* 1989, 33, 79- 89. doi: 10.1002/bit.260330111.

Michael Rajesh, A., Bhatt, S. A., Brahmbhatt, H., Anand, P.S., Popat, K.M., Taste masking of ciprofloxacin by ion-exchange resin and sustain release at gastric-intestinal through interpenetrating polymer network. *Asian J. Pharm. Sci.*, 10, 331, 2014.

Mollah M. Z. I., Zahid H. M., Mahal Z., Faruque M. R. I. and Khandaker M. U. (2021) The Usages and Potential Uses of Alginate for Healthcare Applications. *Front. Mol. Biosci.* 8:719972. doi: 10.3389/fmolb.2021.719972.

Mukesh Shinde B, Ganesh Dama. Studies on development of *in situ* gelling system of ciprofloxacin. *Journal of Advanced Drug Delivery*. 2014;1(1),2. http://www.jadd.in/Content/Paper/20191957531000011201919575310000001.pdf.

Mythri, G., Kavitha, K., Kumar, M. R., Jagadeesh Singh, S. D., Novel mucoadhesive polymers—A review. *J. Appl. Pharm. Sci.*, 1, 8, 37, 2011.

Nause R. G., Reddy R. D., Soh, J. L. P. "Propylene glycol alginate," in *Handbook of Pharmaceutical Excipients*, pp. 594–595, Pharmaceutical Press, London, UK, 6th edition, 2009.

Nayak A. K., Pal. D. Development of pH-sensitive tamarind seed polysaccharide-alginate composite beads for controlled diclofenac sodium delivery using response surface methodology. *International Journal of Biological Macromolecules*. 2011; 49(4):784. doi: 10.1016/j.ijbiomac.2011.07.013.

Nayak P. L., Debasish Sahoo. Chitosan-alginate composites blended with Cloisite 30B as a novel drug delivery system for anticancer drug paclitaxel. *International Journal of Plastics Technology*. 2011; 15(1): 68. doi:10.1007/s12588-011-9000-6.

Niekraszewicz, B. and Niekraszewicz, A., The structure of alginate, chitin and CS fibres, in: *Handbook of Textile Fibre Structure*, p. 266, Woodhead Publishing Limited, Cambridge, UK, 2009.

Okunlola A., Patel R. P, Odeku O. A. Evaluation of freeze-dried pregelatinized Chinese yam (Dioscorea oppositifolia) starch as a polymer in floating gastroretentive metformin microbeads. *Journal of Drug Delivery Science and Technology*. 2010; 20 (6): 457-465. doi.org/10.1016/S1773-2247(10)50079-0.

Otterlei M., Ostgaard K., Skjakbraek G., Smidsrod O., Soonshiong P., Espevik T. Induction of cytokine production from human monocytes stimulated with alginate. J Immunother, 1991; 10: 286-291. doi:10.1016/j.progpolymsci.2011.06.003.

Oya Şanlı, Nuran Ay, Nuran Işıklan. Release characteristics of diclofenac sodium from poly (vinyl alcohol)/sodium alginate and poly(vinyl alcohol)-grafted-poly(acrylamide)/sodium alginate blend beads. *European Journal of Pharmaceutics and Biopharmaceutics*. 2007; 65(2): 204. doi: 10.1016/j.ejpb.2006.08.004.

Pelletier S., Hubert P., Payan E., Marchal P., Choplin L., Dellacherie E. Amphiphilic derivatives of sodium alginate and hyaluronate for cartilage repair: Rheological properties. *J Biomed Mater Res.* 2001; 54:102-108. doi: 10.1002/1097-4636(200101)54:1<102::aid-jbm12>3.0.co;2-1.

Peppas N. A., Buri P. A. Surface interfacial and molecular aspects of polymer bioadhesion on soft tissues. *J. Control Release*. 1985; 2: 257-275. doi.org/10.1016/0168-3659(85)90050-1.

Peteiro C. (2017). Alginate Production from marine Macroalgae, with Emphasis on Kelp Farming. *Alginates Their Biomed. Applic.*, 27–66. doi:10.1007/978-981- 10-6910-9-2.

Raghav S., Jain, P, Kumar D. *Alginates: Properties and Applications*.

Rajesh N., Siddaramaiah, Gowda D. V., Somashekar C. N. Formulation and evaluation of biopolymer based transdermal drug delivery, *International Journal of Pharmacy Pharmaceutical Sciences,* 2010, 2, Suppl 2:142-148. https://innovareacademics.in/journal/ijpps/Vol2Suppl2/569.pdf.

Ramesh Babu V., Malladi Sairam, Kallappa M., Hosamani, Tejraj Aminabhavi M.. Preparation of sodium alginate–methylcellulose blend microspheres for controlled release of Nifedipine. *Carbohydrate Polymers.* 2007; 69(2): 241-250. doi.org/10.1016/j.carbpol.2006.09.027.

Reddy S. G., Alginates - A Seaweed Product: Its Properties and Applications. *Properties and Applications of Alginates*. Edited by Irem Deniz, Esra Imamoglu and Tugba Keskin-Gundogdu.

Reddy, S. G., Pandit, A. S. and Thakur, A. *Study on the Effects of crosslink agents on Sodium Alginate and Lignosulphonic Acid Blends Polymer* (Korea), Vol. 40, No. 1, pp. 63-69 (2016). doi:10.7317/pk.2016.40.1.63.

Remminghorst, U. and Rehm, B. H. A., Bacterial alginates: From biosynthesis to applications. N*Biotechnol. Lett.*, 28, 1701, 2006.

Repka M. A., and Singh, S. "Alginic acid," in *Handbook of Pharmaceutical Excipients*, pp. 20–22, Pharmaceutical Press, London, UK, 6th edition, 2009.

Rinaudo M. On the abnormal exponents' aη and aD in mark-Houwink type equations for wormlike chain polysaccharides. *Polym Bull.* 1992; 27:585-589. doi.org/10.1007/BF00300608.

Rohit Chaudhary, M. D. Shamim Qureshi, Jitendra Patel, Uttam Prasad Panigrahi, Giri, I. C. Formulation, Development and *in-vitro* evaluation of mucoadhesive buccal patches of methotrexate. *International Journal of Pharma Sciences and Research.* 2010; 1(9):357.

Serp D., Cantana E., Heinzen C., Stockar U. V., Marison I. W. Characterization of an encapsulation device for the production of monodisperse alginate beads for cell immobilization. *Biotech. Bioeng.* 2000; 70 (1):41-53. doi.org/10.1002/1097-0290(20001005)70:1<41::AID-BIT6>3.0.CO;2-U.

Shahand S. A., Thassu, D. "Ammoniumalginate," in *Handbook of Pharmaceutical Excipients*, p. 41, Pharmaceutical Press, London, UK, 6th edition, 2009.

Shamkhani A., Duncan R. Radioiodination of alginate via covalently-bound tyrosinamide allows monitoring of its fate *in vivo*. *J Bioact Compat Polym* 1995;10:4-13.doi.org/10.1177/088391159501000102.

Shimokawa T., Yoshida S., Takeuchi T., Murata K., Ishii T., Kusakabe I. (1996) Preparation of two series of oligo-guluronic acids from sodium alginate by acid hydrolysis and enzymatic degradation. *Biosci. Biotechnol. Biochem.*, 60, 1532.

Silva E. A., Mooney D. J. Effects of VEGF temporal and spatial presentation on angiogenesis. *Biomaterials* 2010; 31: 1235-1241. doi: 10.1016/j.biomaterials.2009.10.052.

Siqueira, P., Siqueira, É., De Lima, A. E., Siqueira, G., Pinzón-Garcia, A. D., Lopes, A. P., et al. (2019). Three-dimensional Stable Alginate-Nanocellulose Gels for Biomedical Applications: towards Tunable Mechanical Properties and Cell Growing. *Nanomaterials* 9 (1), 78. doi:10.3390/nano9010078.

Smeds K. A., Grinstaff M. W. Photocrosslinkable polysaccharides for *in situ* hydrogel formation. *J. Biomed. Mat. Res.* 2001, 54 (1),115-12 1. doi.org/10.1002/1097-4636(200105)55:2<254::AIDJBM1012> 3.0.CO;2-5.

Smidsrod O., Skjak-Bræk G. Alginate as immobilization matrix for cells. *Trend Biotechnol.*1990;8:71-78. doi.org/10.1016/0167-7799(90)90139-O.

Soares P. J., Santos J. E., Chierice G. O., Cavalheiro E. T. G. Thermal behavior of alginic acid and its sodium salt, Eclet. *Química.* 2004, .29 (2): 57-64. doi.org/10.1590/S0100-46702004000200009.

Stabler C., Wilks K., Sambanis A., Constantinidis. The effects of alginate composition on encapsulated bTC3 cells. *Biomaterials* 2001, 22 (11), 1301-1310. doi.org/10.1016/S0142- 9612(00)00282-9.

Stevens Molly, M., Qanadilo Hala, F., Langer, R., Prasad Shastri, V., A rapid-curing alginate gel system: Utility in periosteum-derived cartilage tissue engineering. *Biomaterials*, 25, 887, 2004.

Sun, J. and Tan, H., Alginate-based biomaterials for regenerative medicine applications. *Materials*, 6, 1285, 2013.

Suzuki Y., Nishimura Y., Tanihara M., Suzuki K., Nakamura T., Shimizu Y., Yamawaki Y., Kakimaru Y. Evaluation of a novel alginate gel dressing: Cytotoxicity to fibroblasts *in vitro* and foreign-body reaction in pig skin *in vivo J Biomed Mater Res.* 1998; 39:317-22. doi.org/10.1002/(SICI)1097- 4636(199802)39:2<317::AID-JBM20>3.0.CO;2-8.

Szekalska M., Puciłowska A., Szyma´nska E., Ciosek P., Winnicka K. Alginate: Current Use and Future Perspectives in Pharmaceutical and Biomedical Applications. *Int. J. Polym. Sci.* 2016; 8:1-17. doi.org/10.1155/2016/7697031.

ter Horst, B., Moiemen N. S., Grover L. M., and Liam M. G. (2019). *Biomaterials for Skin Repair and Regeneration.* Netherland: Elsevier, 151–192. doi:10.1016/ B978-0-08-102546-8.00006-6.

Thomas, S., Alginate dressings in surgery and wound management—Part 1. *J. Wound Care*, 9, 2, 56, 2000.

Tonnesen, H. H. and Karlsen, J., Alginate in drug delivery systems. *Drug Dev. Ind. Pharm.*, 28, 621, 2002.

Torres M. L., Fernandez J. M., Dellatorre F. G., Cortizo A. M., Oberti T. G. Purification of alginate improves its biocompatibility and eliminates cytotoxicity in matrix for bone tissue engineering. *Algal Research,* 2019;40:101499, doi.org/10.1016/j.algal.2019.101499.

Vandenbossche, G. M. R., Remon, J. P. Influence of the sterilization process on alginate dispersions. *J. Pharm. Pharmacol.* 1993, 45 (16),484-486. https://doi.org/10.1111/j.2042-7158.1993.tb05582.x/

Wahl, E. A., Fierro, F. A., Peavy, T. R., Hopfner, U., Dye, J. F., Machens, H. G., Egaña, J. T., & Schenck, T. L. *In vitro* evaluation of scaffolds for the delivery of mesenchymal stem cells to wounds. *Biomed. Res. Int.*, 108571, 14, 2015.

Wang, W., Zhou, S., Sun, L., Huang, C., Controlled delivery of paracetamol and protein at different stages from core–shell biodegradable microspheres. *Carbohydr. Polym.*, 79, 437, 2010.

Wang, X., Xie, Z. P., Huang, Y., Cheng, Y. B., Gelcasting of silicon carbide based on gelation of sodium alginate. *Ceram. Int.*, 28, 865, 2002.

Wells L. A., Sheardown H. Extended release of high pI proteins from alginate microspheres via a novel encapsulation technique. *Eur J Pharm Biopharm* 2007;65: 329-335. doi:10.1016/j.ejpb.2006.10.018.

Yan, H., Yonghui, L., Qiongyu, Z., Haiyan, L., Jinliang, P., Yuhong, X., et al. (2017). No Title. *J. Mat. Chem. B*. 5, 3315–3326.

Yao B. L., Ni C. H., Xiong C., Zhu C. P., Huang B. Hydrophobic modification of sodium alginate and its application in drug controlled release. *Bioprocess Biosyst Eng.* 2010; 33:457-463. doi: 10.1007/s00449-009-0349-2.

Yu C. Y., Yin B. C., Zhang W., Cheng S. X., Zhang X. Z., Zhuo R. X. Composite microparticle drug delivery systems based on chitosan, alginate and pectin with improved pH-sensitive drug release property," *Colloids and Surfaces B: Biointerfaces*, 2009; 68(2):245-249. doi: 10.1016/j.colsurfb.2008.10.013.

Zeng W. M. Oral controlled release formulation for highly water-soluble drugs: Drug–sodium Alginate–xanthan gum–zinc acetate matrix. *Drug Development and Industrial Pharmacy*. 2004; 30 (5):491-495. doi.org/10.1081/DDC-120037479.

Zeng, Q., Han, Y., Li, H., and Chang, J. (2015). Design of a Thermosensitive Bioglass/agarose-Alginate Composite Hydrogel for Chronic Wound Healing. *J. Mater. Chem.* B 3 (45), 8856–8864. doi:10.1039/c5tb01758k.

Zhang, W. and He, X., Microencapsulating and banking living cells for cell-based medicine. *J. Healthc. Eng.*, 2, 427, 2011.

Zhang, W. and He, X., Microencapsulating and banking living cells for cell-based medicine. *J. Healthc. Eng.*, 2, 427, 2011.

Zhao X. H., Huebsch N., Mooney D. J., Suo Z. G. Stress-relaxation behavior in gels with ionic and covalent crosslinks. *J Appl Phys.* 2010; 107 063509/1-063509/5. doi: 10.1063/1.3343265.

Zhao, Y., Shen, W., Chen, Z., Wu, T., (2016) Freeze-thaw induced gelation of alginates. *Carbohydr. Polym.*, 148, 45.

Zhou, Q., Kang, H., Bielec, M., Wu, X., Cheng, Q., Wei, W., & Dai, H. (2018). Influence of Different Divalent Ions Cross-Linking Sodium Alginate-Polyacrylamide Hydrogels on Antibacterial Properties and Wound Healing. *Carbohydr. Polym.* 197, 292–304.

Zimmermann U., Klock G., Federlin K., Haning K., Kowaslski M., Bretzel R. G., Horcher A., Entenmann H., Siebers U., Zekorn T. Production of mitogen contamination free alginates with variable ratios of mannuronic to guluronic acid by free flow electrophoresis. *Electrophoresis*. 1992; 13:269-74. doi.org/10.1002/elps.1150130156.

Chapter 4

Alginate Hydrogel Beads as Building Blocks for Delivering Polyphenols

Ina Ćorković[1], Drago Šubarić[1], Jurislav Babić[1], Anita Pichler[1], Josip Šimunović[2] and Mirela Kopjar[1,*]

[1]University of Josip Juraj Strossmayer in Osijek, Faculty of Food Technology Osijek, Osijek, Croatia
[2]Department of Food, Bioprocessing and Nutrition Sciences, North Carolina State University, Raleigh, NC, USA

Abstract

Alginate is a biopolymer which, due to its biocompatible and biodegradable properties has numerous potential applications in various food industry sectors. Development of alginate hydrogel beads by the process of microencapsulation enables the preservation of the entrapped material. As encapsulated materials polyphenols were selected. Polyphenols are well-known health-promoting components which are increasingly explored. Through the improvement of their stability, their range of applications is expanding. The purpose of this chapter was to build upon the available literature data concerning polyphenols as encapsulated materials in alginate hydrogel beads. The main challenges concerning the alginate porous structure were addressed as well. Considering the versatile properties of polyphenols and alginate, these components certainly deserve the attention of the scientific community in order to be fully exploited.

Keywords: alginate, beads, polyphenols

---

[*] Corresponding Author's Email: mirela.kopjar@ptfos.hr.

In: Properties and Applications of Alginate
Editor: Michael Y. Wilkerson
ISBN: 979-8-88697-371-6
© 2022 Nova Science Publishers, Inc.

Introduction

Since eating habits are rapidly changing, the food industry is facing demands for the development of more nutritious, safe to eat, natural, less processed and interactive foods. Development of alginate hydrogel beads serves as a potential tool by which these challenges can be met. Microencapsulation technology uses probiotics, certain vitamins, antioxidants and other bioactives that benefit human health and utilizes them to form gel systems as alginate hydrogel beads. Since hydrogel particles are able to preserve the beneficial food structures and nutrients, health-promoting bioactives can be incorporated into an array of food products (Perez and Gaonkar, 2015). Products of this process are microcapsules and microspheres, where microcapsules contain the active compounds in the core of the hydrogel bead and microspheres have the encapsulated material dispersed and trapped throughout the entire hydrogel bead's matrix (Lupo et al., 2013). Microencapsulation is recognized as a method that improves delivery systems by prolonging or controlling the release of the encapsulant and targeting of the bioactives. Polyphenols are commonly used as functional ingredients to enhance the nutritional value of foods and beverages. These bioactive compounds prevent or retard the development of various diseases (Pasukamonset, Kwon and Adisakwattana, 2016). Characteristics of effective delivery systems are high encapsulation efficiency, chemical stability and controlled release (Yun, Devahastin and Chiewchan, 2021). Biopolymers such as chitin, chitosan, glucan, cellulose, agar and alginic acid are isolated from algae, fungi and organisms and possess antioxidant, anticancer, antimicrobial and anti-inflammatory bioactivities. Among materials mentioned, sodium alginate stands out for its biodegradable and biocompatible properties and thus it has recently gained attention in the development of drug delivery systems. If dissolved in water, alginate has non-Newtonian consistency, while it is insoluble in ethanol and ether (Aldawsari et al., 2021). In addition to all the mentioned benefits, there are certain issues about the application of alginate beads in the food industry such as their high porosity, which may cause an uncontrolled release of the encapsulated material or instability of beads as a result of Ca^{2+} leaching after salt exchange and complexation with other molecules. Concentration of alginate solution is limited to the values that produce spherical and well-shaped beads because alginate solutions have high viscosity, even at low concentrations. The formed gel network is often not a sufficient barrier for the encapsulated material and there is a growing effort to improve these unsatisfactory properties of the beads by combining the alginate with other biopolymers (Atencio et al., 2020).

Unlike other recently published reviews dealing with the utilization of alginate (Li, Wei and Xue, 2021; Senturk Parreidt et al., 2018; Sneha Nair et al., 2020), in this chapter we focused on the incorporation of polyphenols into alginate systems. Taking into consideration all of the previously mentioned, the objective of the current manuscript was to review the possibility of using hydrogel beads prepared from alginate and alginate combined with other biopolymers as wall materials for microencapsulation of polyphenols from different plant materials.

Alginate – Physicochemical Properties and Sources

Alginate is a water-soluble polysaccharide consisting of 1 -> 4 linked α-L-guluronic acid (G) and β-D-mannuronic acid (M) (Brownlee et al., 2005). The proportion and distribution of these segments affect their chemical and physical properties. Length and number of guluronate (G) segments determine the affinity of alginate towards multivalent cations such as Ca^{2+} which leads to the formation of ionically cross-linked gels (Davidovich-Pinhas and Bianco-Peled, 2010), while MG segments influence the flexibility of polysaccharide chains and reduce the viscosity of alginate solution (Brownlee et al., 2005). Due to the possession of such heteropolymeric structure, polymer sections do not allow uniform packing to arise and in this way, alginate properties are a result of the MG ratio (Smith and Senior, 2021). Furthermore, the alginate structure depends on the alginate source. For instance, the stipe and outer cortex of brown alga *Laminaria hyperborea* for the most part consist of guluronic acid, while leaves consist of mannuronic acid. Higher amounts of mannuronic acid are present in the fruit of *Ascophyllum nodosum* and old tissue contains higher amounts of guluronic acid (Puscaselu et al., 2020). Mechanical properties and degradation rates are determined by the molecular weight of alginate. Higher molecular weights decrease the number of reactive positions available for hydrolytic degradation and cause a slower degradation rate (Sun and Tun, 2013). Molecular weight of alginate ranges between 33 000 and 400 000 g/mol and it is usually expressed as an average of all the molecules present in the sample (Szekalska et al., 2016). Higher pH values cause higher swelling ratios as a result of chain expansion from the presence of ionic carboxylate groups on the backbone (Sun and Tun, 2013). Chemical modification of alginate with propylene glycol allows its application under low pH values and binding of chunked, milled or flaked food products (Augst, Kong and Mooney, 2006). The structure of alginate was explained in detail in numerous previously published studies (Ching, Bansal and Bhandari, 2017; Cao et al., 2020; Sanchez-Ballester, Bataille and Soulairol, 2021).

Gelation Properties

Dimerization of G segments is responsible for the gelation of alginate. After the addition of Ca^{2+} ions, two G chains from the opposite sides are bonded and consequently, a diamond-shaped hole is formed. This hole contains a hydrophilic cavity that binds Ca^{2+} ions using the oxygen atoms from the carboxyl groups. As a result of the junction zone, the "egg-box" shape is made. To achieve this 3-D network, for each cation 4 G segments are necessary. However, to form stable structures 8 to 20 surrounding G residues are required (Ching, Bansal and Bhandari, 2017). The process of the formation of egg-box junctions occurs following the "zipper" mechanism which describes the first binding of the cross-linking ions as thermodynamically less beneficial as opposed to the subsequent binding (Skjåk-Bræk, Donati and Paoletti, 2015). Once the gelation had occurred, molecules of water that are entrapped within the alginate matrix are still free to migrate and this opens a wide range of possibilities for utilization in encapsulation techniques (Tønnesen and Karlsen, 2002). In the food industry, the application of other cations such as Pb^{2+}, Cu^{2+} and Sr^{2+} is limited due to their toxic characteristics (Ching, Bansal and Bhandari, 2017).

External gelation is a method typically applied for the production of alginate particles and it involves dripping of the polymer solution into the cross-linking or hardening solution. On the other hand, inverse gelation describes the process when $CaCl_2$ solution is added into alginate solution. The inverse gelation is usually applied to produce hydrogel beads filled with lipophilic bioactives or for the combination of compounds with different solubility (Celli et al., 2016).

Slow gelation of alginate ensures the creation of gels with mechanical integrity (Szekalska et al., 2016). Besides cross-linking, gelation of alginate can be achieved by applying the phase transition, cell cross-linking, free radical polymerization and click reactions (Sun and Tun, 2013).

Alginate Sources

Alginate constitutes on average 40% of the dry matter of brown algae (*Phaeophyceae*) including *Laminaria hyperborean, Laminaria japonica, Laminaria digitata, Macrocysric pyrifera* and *Ascophyllum nodosum* species (Lee and Mooney, 2012). It appears in the cell walls of these organisms as a gel containing different ions (e.g., calcium, sodium, magnesium, strontium and barium) (Pereira and Cotas, 2020). In brown algae, alginate has the same role as cellulose and pectin have in terrestrial plants (Skjåk-Bræk, Donati and Paoletti, 2015). In addition to the alginate from algae, bacteria also produce alginate while commercially it is most often extracted from algae biomass (Pereira and Cotas, 2020). The bacteria that produce it are members of genera *Pseudomonas* and *Azotobacter*. Still, alginates isolated from seaweeds have wider applications than those with the bacterial origin and are therefore being studied more extensively (Skjåk-Bræk, Donati and Paoletti, 2015).

Extraction of alginate from algae was patented by E. C. C. Stanford in 1881 who proposed soaking of algae with water or diluted acid followed by extraction with sodium carbonate and precipitation of alginate as a result of acid addition (Pereira and Cotas, 2020). Alginic acid is usually converted at the end of this treatment to a salt and sodium alginate is the form in which it is most commonly found (Tønnesen and Karlsen, 2002).

Alginate Characteristics

Thickening, stabilizing and emulsifying properties of alginate are of extreme importance for the food industry. Its applications are found in syrups, ice cream toppings, sauces, instant milk desserts, fruit pies and bakery cream fillings. Separation of phases in mayonnaise, salad dressings and other water-in-oil emulsions could be prevented by the addition of alginate. It also avoids the formation of ice crystals during freezing. This effect is very important in the ice cream industry when an undesirable crunchy texture appears once the ice cream softens between the supermarket and home freezer and forms ice crystals. In addition to fulfilling the usual requirements in the food industry by applying the novel encapsulation methods, alginate can accomplish delayed-release, thermal protection and desirable sensory properties of entrapped material. Enzymes, organic acids, amino acids, alcohols and bacterial metabolites as well as probiotic cells are materials, which are usually captured within the alginate structure (Qin et al., 2018). Before using alginate in any food applications, it needs to be properly purified, as the risk of contamination from proteins or endotoxins present in crude alginate must be minimized (Tam et al., 2011). Currently, alginate beads are widely used as the imitation forms of various foods such as the artificial fish roe. Sphericity, diameter and rupture strength of prepared beads are important factors affecting the appearance and texture properties of the imitation foods (Jeong et al., 2020).

Besides alginate, there are other hydrocolloids such as plant gums (locust and guar gum), cellulose derivates (methylcellulose and carboxymethylcellulose), marine hydrocolloids (carrageenan and agar) which have similar thickening and gelling properties. Although each hydrocolloid has its specific properties, alginate differs from others by its novel structural and related functional properties (Qin et al., 2018). Among the algal phytocolloids including alginate, carrageenan and agar, alginate is the most extensively produced polysaccharide (Brownlee et al., 2005). According to Puscaselu et al. (2020), there are numerous benefits of alginate such as ease of gelation, nontoxicity, biodegradability, regenarability, natural resources, low cost, biocompatibility and inert nature. Another advantage of this biopolymer is its gelling behavior, which is not harsh and thus the typically unstable compounds are encapsulated with minimal or no damage (Augst, Kong and Mooney, 2006). Retention of moisture is achievable because of the hydrophilic nature of alginate, which improves organoleptic properties of foodstuffs and consequently improves their consumer acceptance (Qin et al., 2018). On the other hand, unpleasant odor, depolymerization at temperatures higher than 60°C and poor stability, are some of the disadvantages that need to be considered when choosing alginate. These obstacles are usually overcome by combining different biopolymers with alginate (Puscaselu et al., 2020).

In addition to the superior physical characteristics that expand its application in different industries, another positive property of alginate is that it is an important source of dietary fiber, which reduces cholesterol and glucose uptake and consequently benefits gastrointestinal and cardiovascular systems (Puscaselu et al., 2020). Further exploration of alginate application will contribute to different science-based innovations which could benefit the expanding range of traditional exploitation of this polysaccharide (Pereira and Cotas, 2020).

Application of Vibrating Technology to Produce Alginate Beads

Factors influencing the selection of the encapsulation method depend on the properties of the active agent and wall materials, process parameters affecting the efficiency, time of production and risk assessment (Dorati et al., 2013). For the food industry, the most important factor is the size of the particles produced since proper particle size selection allows food manufactures to mask certain unwanted characteristics of products such as graininess or powdery (Ching, Bansal and Bhandari, 2017).

Recently, the vibration technique has been studied widely thus the emphasis in this study will be on this particular method. Other techniques to produce alginate beads such as simple dripping/extrusion, modified extrusion (electrostatic atomization, jet cutter method, spinning disk/nozzle, spray nozzle), impinging aerosol gel formation technique, emulsification technique or microfluidics systems, templating method are described in detail in the study of Ching, Bansal and Bhandari (2017).

Encapsulation by extrusion with vibrating technology was recognized as an effective tool for the preservation of bioactive ingredients. Functional foods with incorporated bioactive compounds are developed to improve the bioactivity, stability and bioavailability of unstable but highly valuable components. By applying encapsulation techniques, the degradation of susceptible compounds by environmental factors is reduced. Encapsulator applies vibrational frequency that causes laminar jet break-up and consequently homogeneous-sized and shaped particles are formed (Fangmeier et al., 2019).

The most important parameters which need to be well set and controlled to achieve particles with desirable characteristics are the size of nozzles, jet flow rate, vibration frequency and electrode tension (Dorati et al., 2013). If the liquid jet flow rate is not high enough, the extruded liquid is stuck at the nozzle edge until gravitational force prevails the surface tension and consequently, the drop is released. More droplets are formed because of a small rise in the velocity and its further increase may cause unwanted coalescence of the particles. With increasing velocity, an uninterrupted laminar jet is formed, and is broken up into droplets by vibrations. It can also be broken by the high frictional forces when the jet is sprayed. This technique is usually applied at an industrial scale since it allows production of large amounts of homogenous droplets. If the velocity of the flow is set too high, the impact forces on the droplets will also increase and cause their deformation. This can be avoided by reducing the jet velocity and the distance between the nozzle and the hardening solution. Another option is the addition of surfactants that will not cause droplet deformation and coalescence such as Tween 80. The proper flow rate is set by using the pressure reduction valve. Polymer stream is achieved by one of two possible systems. The first one is ensured by a pulsation-free syringe pump and the second one by a pressure regulation system that supplies the compressed air to a container with polymer solution at a pressure ranging between 0.1 and 2 bar. In comparison to the syringe pump, this system allows the production of higher quantities and uniform droplets of which subsequently beads are formed in a single process (Whelehan and Marison, 2011).

Both nozzle diameter and sinusoidal vibration frequency affect the droplet size. Frequency also regulates the amount of the generated particles, where vibration frequency of 7000 Hz produces 7000 droplets per second. Electrode tension aims to stabilize the droplet jet and disperse it to prevent the formation of droplet clusters (Dorati et al., 2013). The recommendation is not to add potential higher than 2.15 kV since it may cause unstable droplet formation and burst (Whelehan and Marison, 2011). The diameter of particles (d) can be defined by a mathematical equation:

$$d = \sqrt{\frac{6 \cdot V}{\pi \cdot f}} \tag{1}$$

where V represents the flow rate and f the applied vibration frequency (Dorati et al., 2013).

The nozzles are made of stainless steel with the orifices ranging in diameters from 50 to 1000 µm and are an efficient system for the production of particles ranging from 100 to 2000 µm. It is important to set the viscosity of the encapsulation mixture to obtain particles with a satisfactory size. If the viscosity of the mixture is relatively high, the vibration effect will not be sufficient to produce the particles and the possibility to clog the nozzles is higher. On the other hand, if the viscosity is not high enough, produced particles will not have the desirable spherical shape (Fangmeier et al., 2019).

In the study of Nemethova, Lacik and Razga (2014), the restrictive role of viscosity was investigated. It was observed that when experiments were performed within the optimal working range for specific viscosity, obtained particles had desired spherical shapes and uniform sizes, while experiments outside the optimal working range caused increased size dispersity and lack of spherically shaped microparticles.

Buchi, Nisco Engineering AG, Inotech Biotechnologies AG and Brace Technologies produce commercially available encapsulators that apply the vibration technology to prepare hydrogel particles. All these devices are designed to fulfill the research requirements and are not suitable for large-scale production. Therefore, design of the devices that can be applied in the industry processes remains a challenge. However, all these encapsulators consist of main parts with the same performances: (1) feeding pump containing the material which needs to be encapsulated dissolved in the wall material solution; (2) nozzle of a certain orifice diameter for the production of the laminar jet; (3) control system to determine frequency; (4) stroboscopic light that allows the observation of particles as they are being formed; (5) gelation solution which is constantly stirred and allows the polymerization and formation of the particles (Fangmeier et al., 2019).

Device for the production of beads comes in two variants: in sterile and non-sterile conditions. To ensure the sterile surroundings, a glass casing is installed around the device. Another difference in the system configuration is the replacement of the single-flow nozzle with a concentric system. A concentric nozzle system consists of an internal and an external nozzle which is regularly larger than the internal one. Syringe pump, pressure regulation or a combination of both are applied to produce microcapsules using a concentric system. In addition to the already mentioned factor that affects the particle size in a single-nozzle system, the size of beads produced using a concentric nozzle system is affected by the diameter of the outer nozzle. Thickness of the membrane material is a crucial property affecting the retention or the release of the encapsulated material (Whelehan and Marison, 2011).

The main disadvantages of this technique are small production rates of microparticles and the restricted choice of biopolymers which is usually based on the application of alginate alone or in combination with other biopolymers (Whelehan and Marison, 2011).

Microencapsulation of Polyphenols into Alginate Particles

Polyphenols are secondary metabolites present in plant materials such as fruits, vegetables and grains. They are reported to have antioxidant, antibacterial, antimutagenic, anti-inflammatory activities. Due to their derivable molecular structure, polyphenols possess unique chemical and physical properties which enable them to interact with polysaccharides, proteins, alkaloids, metal ions and capture free radicals. Poor stability, low solubility and unpleasant taste restrain their utilization for the development of functional foods (Feng et al., 2020). Polyphenols' ability to scavenge free radicals is reduced by the high temperatures which are usually applied in the food industry and acidic conditions like those present in the human gastrointestinal system (Abdin et al., 2020). Moreover, polyphenols possess characteristic astringent/bitter taste which also limits their incorporation into conventional foods (Díaz-Bandera et al., 2013). Protection of polyphenols from environmental factors and masking of the unpleasant taste can be achieved by the microencapsulation process, i.e., incorporation in the hydrogel beads. However, to conduct effective microencapsulation, it is important to select a suitable wall material. Since alginate has desirable biocompatible, biodegradable and emulsifying properties it is a suitable polymer for the preparation of hydrogels and microcapsules filled with different polyphenols. Besides all the beneficial effects of alginate, it regrettably, also has high porosity, rapid degradation under simulated intestinal fluids and a lack of stability under thermal treatment (Sheng et al., 2021). The gelation process of alginate depends on the copolymer nature. The explanation of the gelation processes of alginate combined with different polymers is described in detail elsewhere (Bennacef et al., 2021). Different proteins and polysaccharides are added to the alginate matrices to improve the porous structure and degradation of alginate (Feltre et al., 2020). Increased stability of the hydrogel beads is achieved by the interactions of alginate with other polysaccharides, while the addition of sucrose ensures the cryo-protective effect (Aguirre Calvo, Perullini and Santagapita, 2018).

Considering all the mentioned advantages and disadvantages of using alginate as a wall material in the preparation of hydrogel beads, Table 1 presents the results from studies that investigated the influence of alginate and various edible polymers on the encapsulation of polyphenols. Additionally, main conclusions of those studies are given in this Table.

Table 1. A review of studies that investigated alginate and other polymers as wall materials for the microencapsulation of polyphenols

Polymer(s) used	Forms	Entrapped material	Observation	Reference
Alginate-pectin	Hydrogel particles	Anthocyanins from purple corn and blueberry extracts	An increase of polysaccharides' concentration caused an increase in encapsulation efficiency.	Guo, Giusti and Kaletunç (2017)
Alginate-starch	Beads	Polyphenols from yerba mate extracts	The addition of starch improved the encapsulation efficiency from 55 to 65%.	López Córdoba, Deladino and Martino (2013)
Alginate-corn starch	Beads	Wheat germ oil containing polyphenols, fatty acids tocopherols, phytosterols and carotenoids	The presence of starch caused better protection from oil oxidation during 6-day storage.	Feltre et al. (2020)
Alginate	Beads	Polyphenols from extracts of *Clitoria ternatea* petal flower	Alginate beads have successfully reduced polyphenol degradation and increased biological activity after gastrointestinal digestion (demonstrated a significant increase in antioxidant activity and α-amylase inhibitory activity).	Pasukamonset, Kwon and Adisakwattana (2016)
Alginate-chitosan	Beads	Glucosyl-hesperidin	Alginate-chitosan beads complexed in $CaCl_2$ for 30 min had the best retention ability of the hesperidin derivate. The same was observed after the 7-day storage of the beads.	Ćorković et al. (2021a)
Alginate-pectin	Beads	Polyphenols from chokeberry juice	The addition of pectin to alginate wall material caused an increase of encapsulated total polyphenols.	Ćorković et al. (2021b)
Alginate	Beads	Apple skin polyphenol extract with the addition of *Lactobacillus acidophilus*	Encapsulation increased the survival of probiotics and caused high values of polyphenol content and antioxidant activity.	Shinde, Sun-Waterhouse and Brooks (2013)

Table 1. (Continued)

Polymer(s) used	Forms	Entrapped material	Observation	Reference
Alginate	Beads	Apigenin	Apigenin-loaded beads were successful carriers of antitumor, antibacterial and antioxidant activities, entrapment of apigenin increased proportionally with the increase of polymer concentration.	Aldawsari et al. (2021)
Alginate	Beads	Anthocyanins and polyphenols from jabuticaba peel extract and propolis extract	The obtained beads had high encapsulation efficiency of total polyphenols (98%) and anthocyanins (89%). The release study showed that beads were resistant at gastric pH (1.2) but their complete disintegration occurred at intestinal pH (7.4).	Dallabona et al. (2020)
Alginate-chitosan	Microgel particles	Polyphenols from the shells of *Juglans regia* L.	Microgel formation prevented the degradation of polyphenols by simulated gastrointestinal digestion.	Feng et al. (2020)
Alginate-chitosan	Beads	Polyphenol rich *Murraya koengii* bark extract	The addition of chitosan caused higher encapsulation efficiencies.	Dadwal et al. (2020)
Alginate-whey protein Alginate-hydroxypropyl methylcellulose Pectin-whey protein Pectin-hydroxypropyl methylcellulose	Microbeads	Dandelion (*Taraxacum officinale* L.) polyphenols	The addition of whey proteins and cellulose derivate caused an increase in encapsulation efficiency of polyphenols.	Belščak-Cvitanović et al. (2016)
Alginate Alginate-tapioca starch	Beads	Chlorogenic acid	Encapsulation efficiency increased with the addition of tapioca starch and the highest release of phenolic acid was observed for formulation prepared in weight ratio alginate to starch: 0.75/0.25.	Lozano-Vazquez et al. (2015)
Alginate Alginate-methyl cellulose Alginate-hydroxypropyl methylcellulose	Microcapsules	Grape seed proanthocyanidin extract	The addition of cellulose derivates caused an increase in encapsulation efficiency.	Sheng et al. (2021)

Polymer(s) used	Forms	Entrapped material	Observation	Reference
Alginate	Microcapsules	*Syzygium cumin* seeds polyphenols	Increased thermal stability of polyphenols and increased preservation of polyphenols in gastric conditions, as well as release of polyphenols in intestinal conditions were observed.	Abdin et al. (2021)
Alginate, Alginate-inulin, Alginate-arabic gum, Alginate-chitosan	Beads	Tea polyphenols	The increase of encapsulation efficiency followed the order: alginate-inulin < alginate < alginate-arabic gum < alginate-chitosan.	Li et al. (2021)
Alginate	Beads	Olive oil enriched with carotenoids from orange peel	The change in antioxidant activity of samples before and after encapsulation was negligible.	Savic Gajic et al. (2021a)
Alginate	Microparticles	Pot marigold (*Calendula officinalis* L.) flowers extract	The antioxidant activity of the extract was preserved under simulated gastrointestinal conditions.	Savic Gajic et al. (2021b)
Alginate-collagen	Microspheres	Tea polyphenols	Encapsulated polyphenols showed good antioxidant activity and better stability at neutral pH (7.4) than in the acidic condition.	Ruan et al. (2020)
Alginate	Beads	Polyphenols from propolis extract	Encapsulation efficiency increased when the alginate concentration was higher.	Keskin, Keskin and Kolayli (2019)
Alginate-sucrose, Alginate-sucrose-guar gum, Alginate-sucrose-arabic gum, Alginate-sucrose-low methoxyl pectin, Alginate-sucrose-high methoxyl pectin	Beads	Polyphenols of beetroot leaf and stem	The highest loading efficiency was observed for systems containing guar gum and the lowest for arabic gum.	Aguirre Calvo, Santagapita and Perullini (2019)
Alginate	Microbeads	Cocoa polyphenols	Encapsulation efficiency was around 60% and a certain amount was lost during beads washing.	Lupo et al. (2013)

Table 1. (Continued)

Polymer(s) used	Forms	Entrapped material	Observation	Reference
Alginate-proteins (whey proteins, bovine serum albumine, calcium caseinate, soy proteins, hemp proteins) Alginate-proteins-chitosan-pectin	Microparticles	Flavan-3-ols and caffeine from green tea extract	The combination of alginate and calcium caseinate or whey proteins retained the highest concentration (up to 80%) of polyphenols and caffeine while chitosan and pectin did not improve encapsulation efficiency.	Belščak-Cvitanović et al. (2015)
Alginate-gelatin	Beads	Polyphenols extracted from *Hibiscus sabdariffa* L. calyx	The release rate of polyphenols was reduced by the addition of alginate coat(s).	Diaz-Bandera et al. (2013)
Alginate-basil seed gum-carrageenan-lactate	Beads	Barberry polyphenols	Samples with the highest amounts of alginate had the best antioxidant properties after the storage period.	Maleki et al. (2020)
Alginate	Beads	Polyphenols from pomegranate peels extract	The highest encapsulation efficiency (43.90%) of polyphenols and proanthocyanidins (46.34%) was observed for the beads with 1% extract, 3% alginate that was cured in 0.05 M $CaCl_2$ for 20 min and kept in the gelling bath for 15 min.	Zam et al. (2014)
Alginate	Beads	Chokeberry phenolic extract	For the optimum encapsulation efficiency (94.2%) required conditions were: 2% alginate, 2.5% $CaCl_2$ and 20% of extract.	Tzatsi and Goula (2021)
Alginate Alginate-chitosan	Beads	Yerba mate extract rich in polyphenols	Chitosan as an additional layer caused a decrease in encapsulation efficiency.	Deladino et al. (2008)
Alginate	Beads	Polyphenols from rose hips	The addition of chitosan enhanced the encapsulation of polyphenols.	Stoica, Pop and Ion (2013)
Alginate Alginate-chitosan	Beads	Polyphenols from lemon balm extract	The higher concentrations of alginate created a network with a ticker membrane that inhibited the loss of encapsulated material to the environment.	Najafi-Soulari, Shekarchizadehand Kadivar (2016)

Polymer(s) used	Forms	Entrapped material	Observation	Reference
Alginate	Beads	Grape skin polyphenols	Encapsulation efficiency was 68% and it was affected by liquid-liquid diffusion that occurs in the hardening solution before gelling.	Lavelli and Sri Harsha (2019)
Alginate Alginate-whey protein isolate Alginate-cocoa Alginate-carob powder	Beads	Extract from dandelion polyphenols	The addition of fillers improved the encapsulation of polyphenols by improving a very porous alginate structure.	Bušić et al. (2018)
Alginate Alginate-chitosan	Beads	Grape seed polyphenols from winemaking byproducts	Encapsulation efficiency was up to 92% and this process prevented interactions of proanthocyanidins with different food matrices.	Pedrali, Barbarito and Lavelli (2020)
Alginate	Beads	*Vitis vinifera* grape wastes	Higher values of encapsulation efficiencies related to larger sizes of nozzles and were independent of the extract concentration.	Aizpurua-Olaizola et al. (2015)
Alginate	Beads	Polyphenols from beta Vulgaris cv. beet greens	The direct relationship between mechanical characteristics, swelling of the beads and release rates of polyphenols was observed within *in-vitro* conditions.	Gorbunova et al. (2018)
Alginate-chocolate beads	Beads	*Moringa oleifera* extract rich in phenolic acids and flavonoids	The addition of extract affected the color and shape of the beads: an increase in its concentration caused higher redness, yellowness and chroma and beads were less elliptical. The lowest concentration of polyphenols was observed for beads prepared with 2% of extract and a hardening time of 20 min.	Kaltsa et al. (2021)
Alginate Alginate-medium molecular weight chitosan Alginate-low molecular weight chitosan	Beads	Roselle anthocyanins	With the increasing alginate content up to 3% and anthocyanin concentration in the soaking solution, total polyphenols increased and caused higher antioxidant activities. Low molecular weight chitosan restrained the diffusion of anthocyanins into model acidic beverages.	Nguyen et al. (2022)

Table 1. (Continued)

Polymer(s) used	Forms	Entrapped material	Observation	Reference
Alginate Alginate-hydroxypropyl methylcellulose (HPMC)	Beads	Avocado oil enriched with phloridizin	Alginate beads were bigger and stronger than alginate-HPMC beads. The addition of phloridizin caused improved oxidative stability.	Sun-Waterhouse et al. (2012)
Alginate Chitosan	Beads	Anthocyanins from *Pinot Noir* grape skins	The highest adsorption capacity of chitosan beads was reached at pH 8, while for alginate it was reached at pH 4. Since they interact with different molecular structures of anthocyanins, both alginate and chitosan are good carriers of anthocyanins.	Pinheiro et al. (2021)
Alginate-chitosan	Beads	Mulberry anthocyanins	The highest encapsulation efficiency was observed for alginate beads that were complexed with 0.05% chitosan solution.	Kanokpanont, Yamdech and Aramwit (2018)

Future Perspectives

Microencapsulation has been proposed as a potential method for the preservation of polyphenols from plant sources and a review of the relevant literature confirmed that alginate hydrogel beads can be successfully used as carriers of these valuable bioactive compounds. The obstacles concerning the porous alginate structure can be overcome by the addition of other biopolymers into the wall material matrix. These findings can help solve the formidable task of preserving polyphenols during various processes in the food industry. However, formulation of the optimal encapsulation mixtures will remain the main challenge in order to obtain the hydrogel beads with satisfactory encapsulation efficiency of polyphenols as well as their release. Vibration technology to produce hydrogel beads was found to be reproducible, simple and rapid, although there are also other methods that can be applied for the production of hydrogel beads.

Future experimental studies should be based on improvements of physical properties of alginate hydrogels according to the required applications. This future trend is already in progress, and the exploitation of alginates has been improved from basic to more complex applications such as the targeted delivery and controlled release of the entrapped materials in the food industry.

Acknowledgments

This work was supported by the Croatian Science Foundation under project (IP-2019-04-5749) "Design, fabrication and testing of biopolymer gels as delivery systems for bioactive and volatile compounds in innovative functional foods (bioACTIVEgels)," and Young Researchers' Career Development Project—Training New Doctoral Students (DOK-2020-01-4205).

References

Abdin, M., Salama, M. A., Riaz, A., Akhtar, H. M. S. & Elsanat, S. Y. (2021). Enhanced the entrapment and controlled release of *Syzygium cumini* seeds polyphenols by modifying the surface and internal organization of Alginate-based microcapsules. *Journal of Food Processing and Preservation, 45,* 15100. http://doi.org/10.1111/jfpp.15100.

Aguirre Calvo, T. R., Perullini, M. & Santagapita, P. R. (2018). Encapsulation of betacanins and polyphenols extracted from leaves and stems of beetroot in Ca(II)-alginate beads: A structural study. *Journal of Food Engineering, 235,* 32-40. https://doi.org/10.1016/j.jfoodeng.2018.04.015.

Aguirre Calvo, T. R., Santagapita, P. R. & Perullini, M. (2019). Functional and structural effects of hydrocolloids on Ca(II)-alginate beads containing bioactive compounds extracted from beetroot. *LWT – Food Science and Technology, 111,* 520-526. http://doi.org/10.1016/j.lwt.2019.05.047.

Aizpurua-Olaizola, O., Navarro, P., Vallejo, A., Olivares, M., Etxebarria, N. & Usobiaga, A. (2015). Microencapsulation and storage stability of polyphenols from *Vitis vinifera* grape wastes. *Food Chemistry, 190,* 614-621. http://dx.doi.org/10.1016/j.foodchem.2015.05.117.

Aldawsari, M. F., Ahmed, M. M., Fatima, F., Anwer, M. K., Katakam, P. & Khan, A. (2021). Development and Characterization of Calcium-Alginate Beads of Apigenin: In Vitro Antitumor, Antibacterial, and Antioxidant Activities. *Marine Drugs, 19,* 467. https://doi.org/10.3390/md19080467.

Atencio, S., Maestro, A., Santamaría, E., Gutiérrez, J. M. & González, C. (2020). Encapsulation of ginger oil in alginate-based shell materials. *Food Bioscience, 37,* 100714. https://doi.org/10.1016/j.fbio.2020.100714.

Augst, A., Kong, H. J. & Mooney, D. J. (2006). Alginate Hydrogels as Biomaterials. *Macromolecular Bioscience, 6,* 623-633. https://doi.org/10.1002/mabi.200600069.

Belščak-Cvitanović, A., Bušić, A, Barišić, L., Vrsaljko, D., Karlović, S., Špoljarić, I., Vojvodić, A., Mršić, G. & Komes, D. (2016). Emulsion templated microencapsulation of dandelion (*Taraxacum officinale* L.) polyphenols and ß-carotene by ionotropic gelation of alginate and pectin. *Food Hydrocolloids, 57,* 139-152. http://doi.org/10.1016/j.foodhyd.2016.01.020.

Belščak-Cvitanović, A., Đorđević, V., Karlović, S., Pavlović, V., Komes, D., Ježek, D., Bugarski, B. & Nedović, V. (2015). Protein-reinforced and chitosan-pectin coated alginate microparticles for delivery of flavan-3-ol antioxidants and caffeine from green tea extract. *Food Hydrocolloids, 51,* 361-374. https://doi.org/10.1016/j.foodhyd.2015.05.039.

Bennacef, C., Desobry-Banon, S., Probst, L. & Desobry, S. (2021). Advances on alginate use for spherification to encapsulate biomolecules. *Food Hydrocolloids, 118,* 106782. https://doi.org/10.1016/j.foodhyd.2021.106782.

Brownlee, I. A., Allen, A., Pearson, J. P., Dettmar, P. W., Havler, M. E., Atherton, M. R. & Onsøyen, E. (2005). Alginate as a source of dietary fiber. *Critical Reviews in Food Science and Nutrition, 45,* 497-510. http://doi.org/10.1080/10408390500285673.

Bušić, A., Belščak-Cvitanović, A., Vojvodić Cebin, A., Karlović, S., Kovač, V., Špoljarić, I., Mršić, G. & Komes, D. (2017). Structuring new alginate network aimed for delivery of dandelion (*Taraxacum officinale* L.) polyphenols using ionic gelation and new filler materials. *Food Research International, 111,* 244-255. http://doi.org/10.1016/j.foodres.2018.05.034.

Cao, L., Lu, W., Mata, A., Nishinari, K. & Fang, Y. (2020). Egg-box model-based gelation of alginate and pectin: A review. *Carbohydrate Polymers*, 116389. http://doi.org/10.1016/j.carbpol.2020.116389.

Celli, G. B., Teixeira, A.G., Duke, T. G. & Su-Ling Brooks, M. (2016). Encapsulation of lycopene from watermelon in calcium alginate microparticles using and optimized inverse-gelation method by response surface methodology. *International Journal of Food Science and Technology*, *51*, 1523-1529. http://doi.org/10.1111/ijfs.13114.

Ching, S. H., Bansal, N. & Bhandari, B. (2017). Alginate gel particles–A review of production techniques and physical properties. *Critical Reviews in Food Science and Nutrition*, *57*, 1133-1152. http://doi.org/10.1080/10408398.2014.965773.

Ćorković, I., Pichler, A., Šimunović, J. & Kopjar. M. (2021a). Microencapsulation of glucosyl-hesperidin in alginate/chitosan hydrogel beads. *Food in Health and Disease, scientific-professional journal of nutrition and dietetics*, *10*(2), 39-44.

Ćorković, I., Pichler, A., Ivić, I., Šimunović, J. & Kopjar, M. (2021b). Microencapsulation of Chokeberry Polyphenols and Volatiles: Application of Alginate and Pectin as Wall Materials. *Gels, 7*, 231. https://doi.org/10. 3390/gels7040231.

Dadwal, V., Bhatt, S., Joshi, R. & Gupta, M. (2020). Development and characterization of controlled released polyphenol rich micro-encapsulate of *Murraya koenigii* bark extract. *Journal of Food Processing and Preservation, 44*, 14438. https://doi.org/ 10.1111/jfpp.14438.

Dallabona, I. D., Goetten de Lima, G., Cestaro, B. I., de Souza Tasso, I., Paiva, T. S., Gbur Laureanti, E. J., de Matos Jorge, L. M., Gonçalves da Silva, B. J., Helm, C. V., Mathias, A. L. & Matos Jorge, R. M. (2020). Development of alginate beads with encapsulated jabuticaba peel and propolis extracts to achieve a new natural colorant antioxidant additive. *International Journal of Biological Macromolecules, 163*, 1421-1432. https://doi.org/10.1016/j.ijbiomac.2020.07.256.

Davidovich-Pinhas, M. & Bianco-Peled, H. (2010). A quantitative analysis of alginate swelling. *Carbohydrate Polymers, 79*, 1020-1027. https://doi.org/10.1016/j.carbpol. 2009.10.036.

Deladino, L., Anbinder, P. S., Navarro, A. S. & Martino, M. N. (2008). Encapsulation of natural antioxidants extracted from *Ilex paraguariensis*. *Carbohydrate Polymers, 71*, 126-134. http://doi.org/10.1016/j.carbpol.

Díaz-Bandera, D., Villanueva-Carvajal, A., Dublán-García, O., Quintero-Salazar, B. & Dominguez-Lopez, A. (2013). Release kinetics of antioxidant compounds from *Hibiscus sabdariffa* L. encapsulated in gelatin beads and coated with sodium alginate. *International Journal of Food Science, 48*, 2150-2158. https://doi.org/10. 1111/ijfs.12199.

Dorati, R., Genta, I., Modena, T. & Conti, B. (2013). Microencapsulation of hydrophilic model molecule through vibration nozzle and emulsion phase inversion technologies. *Journal of Microencapsulation*, 1-12. http://doi.org/10.3109/0265 2048.2013.764938.

Fangmeier, M., Lehn, D. N., Maciel, M. J. & de Souza, C. F. V. (2019). Encapsulation of Bioactive Ingredients by Extrusion with Vibrating Technology: Advantages and Challenges. *Food and Bioprocess Technology, 12*, 1472–1486. https://doi.org/10. 1007/s11947-019-02326-7.

Feltre, G., Sartori, T., Silva, K. F. C., Dacanal, G. C., Menegalli, F. C. & Hubinger, M. D. (2020). Encapsulation of wheat germ oil in alginate-gelatinized corn starch beads: Physicochemical properties and tocopherols' stability. *Journal of Food Science, 85*, 2124-2133. http://doi.org/10.1111/1750-3841.15316.

Feng, R., Wang, L., Zhou, P., Luo, Z., Li, X. & Gao, L. (2020). Development of the pH responsive chitosan-alginate based microgel for encapsulation of Jughans regia L. polyphenols under simulated gastrointestinal digestion in vitro. *Carbohydrate Polymers, 250*, 116917. https://doi.org/10.1016/j.carbpol.2020.116917.

Gorbunova, N., Bannikova, A., Evteev, A., Evdokimov, I. & Kasapis, S. (2018). Alginate-based encapsulation of extracts from beta Vulgaris cv. beet greens: Stability and controlled release under simulated gastrointestinal conditions. *LWT – Food Science and Technology, 93*, 442-449. https://doi.org/10.1016/j.lwt.2018.03.075.

Guo, J., Giusti, M. & Kaletunç, G. (2017). Encapsulation of purple corn and blueberry extracts in alginate-pectin hydrogel particles: Impact of processing and storage parameters on encapsulation efficiency. *Food Research International, 107*, 414-422. http://doi.org/10.1016/j.foodres.2018.02.035.

Jeong, C., Kim, S., Lee, C., Cho, S. & Kim, S. B. (2020). Changes in the Physical Properties of Calcium Alginate Gel Beads under a Wide Range of Gelation Temperature Conditions. *Foods, 9*, 180. https://doi.org/10.3390/foods9020180.

Kaltsa, O., Alibade, A., Bozinou, E., Markis P. D. & Lalas, I. S. (2021.) Encapsulation of *Moringa oleifera* Extaract in Ca-Alginate Chocolate Beads: Physical and Antioxidant Properties. *Journal of Food Quality*. https://doi.org/10.1155/2021/ 5549873.

Kanokpanont, S., Yamdech, R. & Aramwit, P. (2018). Stability enhancement of mulberry-extracted anthocyanin using alginate/chitosan microencapsulation for food supplement application. *Artificial Cells, Nanomedicine, and Biotechnology, 46*(4), 773-782. https://doi.org/10.1080/21691401.2017.1339050.

Keskin, M., Keskin, Ş., Kolayli, S. (2019). Preparation of alcohol free propolis-alginate microcapsules, characterization and release property. *LWT – Food Science and Technology, 108*, 89-96. https://doi.org/10.1016/j.lwt.2019.03.036.

Lavelli, V. & Sri Harsha, P. S. C. (2019). Microencapsulation of grape skin phenolics for pH controlled release of antiglycation agents. *Food Research International, 119*, 822-828. https://doi.org/10.1016/j.foodres.2018.10.065.

Lee, K. Y. & Mooney, D. J. (2012). Alginate: Properties and biomedical applications. *Progress in Polymer Science, 37*, 106-126. https://doi.org/10.1016/j.progpolymsci.2011.06.003.

Li, Q., Duan, M., Hou, D., Chen, X., Shi, J. & Zhou, W. (2021). Fabrication and characterization of Ca(II)-alginate-based beads combined with different polysaccharides as vehicles for delivery, release and storage of tea polyphenols. *Food Hydrocolloids, 112*, 106274. https://doi.org/10.1016/j.foodhyd.2020.106274.

Li, D., Wei, Z., Xue, C. (2021). Alginate-based delivery systems for food bioactive ingredients: An overview of recent advances and future trends. *Comprehensive Review of Food Science and Food Safety*, 1-25. http://doi.org/10.1111/1541-4337.12840.

López Córdoba, A., Deladino, L. & Martino, M. (2013). Effect of starch filler on calcium-alginate hydrogels loaded with yerba mate antioxidants. *Carbohydrate Polymers, 95*, 315-323. https://doi.org/10.1016/j.carbpol.2013.03.019.

Lozano-Vazquez, G., Lobato-Calleros, C., Escalona-Buendia, H., Chavez, G., Alvarez-Ramirez, J. & Vernon-Carter, E. J. (2015). Effect of the weight ratio of alginate-modified tapioca starch on the physicochemical properties and release kinetics of chlorogenic acid containing beads. *Food Hydrocolloids, 48*, 301-311. https://doi.org/10.1016/j.foodhyd.2015.02.032.

Lupo, B., Maestro, A., Porras, M., Gutiérrez, J. M. & González, C. (2013). Preparation of alginate microspheres by emulsification/internal gelation to encapsulate Cocoa polyphenols. *Food Hydrocolloids, 38*, 56-65. http://doi.org/10.1016/j.foodhyd.2013.11.003.

Maleki, M., Mortazavi, S. A., Yeganehzad, S. & Pedram Nia, A. (2020). Study on liquid core barberry (*Berberis vulgaris*) hydrogel beads based on calcium alginate: Effect of storage on physical and chemical characterizations. *Journal of Food Processing and Preservation, 44*, 14426. http://doi.org/10.1111/jfpp.14426.

Najafi-Soulari, S., Shekarchizadeh, H. & Kadivar, M. Encapsulation optimization of lemon balm antioxidants in calcium alginate hydrogels. *Journal of Biomaterials Science, Polymer Edition, 27*, 1631-1644. http://doi.org/10.1080/09205063.2016.1226042.

Nemethova, V., Lacik, I. & Razga, F. (2014). Vibration Technology for Microencapsulation: The Restrictive Role of Viscosity. *Journal of Bioprocessing and Biotechniques, 5*, 199. http://doi.org/10.4172/2155-9821.1000199.

Nguyen, Q., Nguyen, T., Nguyen, T. & Nguyen, N. (2022). Encapsulation of roselle anthocyanins in blank alginate beads by adsorption and control of anthocyanin release in beverage by coatings with different molecular weight chitosan. *Journal of Food Processing and Preservation.* http://doi.org/10.1111/jfpp.16438.

Pasukamonset, P., Kwon, O. & Adisakwattana, S. (2016). Alginate-based encapsulation of polyphenols from *Clitoria ternatea* petal flower extract enhances stability and biological activity under simulated gastrointestinal conditions. *Food Hydrocolloids, 61*, 772-779. https://doi.org/10.1016/j.foodhyd.2016.06.039.

Pedrali, D., Barbarito, S. & Lavelli, V. (2020). Encapsulation of grape seed phenolics from winemaking byproducts in hydrogel microbeads – Impact of food matrix and processing on the inhibitory activity towards α-glucosidase. *LWT – Food Science and Technology, 133*, 109952. https://doi.org/10.1016/j.lwt.2020.109952.

Pereira, L. & Cotas, J. (2020.) Alginates – A General Overview. In *Alginates – Recent Uses of This Natural Polymer*, edited by Leonel Pereira. London: IntechOpen. http://doi.org/10.5772/intechopen.88381.

Perez, R. & Gaonkar, A. G. (2014). Commercial Applications of Microencapsulation and Controlled Delivery in Food and Beverage Products. In *Microencapsulation in the Food Industry*, edited by Anilkumar G. Gaonkar, Niraj Vasisht, Atul Ramesh Khare and Robert Sobel. Cambridge, MA: Academic Press. https://doi.org/10.1016/B978-0-12-404568-2.00041-8.

Pinheiro, C. P., Moreira, L. M. K., Alves, S. S., Cadaval Jr, T. R. S. & Pinto, L. A. A. (2021). Anthocyanins concentration by adsorption onto chitosan and alginate beads: Isotherms, kinetics and thermodynamics parameters. *International Journal of Biological Macromolecules, 166*, 934-939. https://doi.org/10.1016/j.ijbiomac.2020.10.250.

Puscaselu, R. G., Lobiuc, A., Dimian, M. & Covasa M. (2020). Alginate: From Food Industry to Biomedical Applications and Management of Metabolic Disorders. *Polymers*, *12*, 2417. http://doi.org/10.3390/polym12102417.

Qin, Y., Jiang, J., Zhao, L., Zhang, J. & Wang, F. (2018). Applications of Alginate as a Functional Food Ingredient. In *Handbook of Food Bioengineering, Biopolymers for Food Design*, edited by Alexandru Mihai Grumezescu and Alina Maria Holban. Cambridge, MA: Academic Press. https://doi.org/10.1016/B978-0-12-811449-0.00013-X.

Ruan, J., Pei, H., Li, T., Wang, H., Li, S. & Zhang, X. (2021). Preparation and antioxidant activity evaluation of tea polyphenol–collagen–alginate microspheres *Journal of Food Processing Preservation*, *45*, 15187. https://doi.org/10.1111/jfpp.15187.

Sanchez-Ballester, N. M., Bataille, B. & Soulairol, I. (2021). Sodium alginate and alginic acid as pharmaceutical excipients for tablet formulation: Structure - function relationship. *Carbohydrate Polymers*, *270*, 118399. https://doi.org/10.1016/j.carbpol.2021.118399.

Savic Gajic, I. M., Savic, I. M., Gajic, D. G., Dosic, A. (2021a). Ultrasound-Assisted Extraction of Carotenoids from Orange Peel Using Olive Oil and Its Encapsulation in Ca-Alginate Beads. *Biomolecules*, *11*, 225. https://doi.org/10.3390/biom11020225.

Savic Gajic, I. M., Savic, I. M., Skrba, M., Dosić, A. & Vujadinovic, D. (2021b). Food additive based on the encapsulated pot marigold (*Calendula officinalis* L.) flowers extract in calcium alginate microparticles. *Journal of Food Processing and Preservation*, 15792. https://doi.org/10.1111/jfpp.15792.

Senturk Parreidt, T., Müller, K., Schmid, M. (2018). Alginate-Based Edible Films and Coatings for Food Packaging Applications. *Foods*, *7*, 170. https://doi.org/10.3390/foods7100170.

Sheng, K., Zhang, G., Kong, X., Wang, J., Mu, W. & Wang, Y. (2021). Encapsulation and characterization of grape seed proanthocyanidin extract using sodium alginate and different cellulose derivatives. *Internatioanl Journal of Food Science and Technology*, *56*, 6420-6430. http://doi.org/10.1111/ijfs.15299.

Shinde, T., Sun-Waterhouse, D. & Brooks, J. (2013). Co-extrusion Encapsulation of Probiotic *Lactobacillus acidophilus* Alone or Together with Apple Skin Polyphenols: An Aqueous and Value-Added Delivery System Using Alginate. *Food Bioprocess Technology*, *7*, 1581-1596. http://doi.org/10.1007/s11947-013-1129-1.

Skjåk-Bræk, G., Donati, I. & Paoletti, S. (2015). Alginate hydrogels: properties and applications. In *Polysaccharide Hydrogels: Characterization and Biomedical Applications*, edited by Pietro Matricardi, Franco Alhaique and Tommasina Coviello. Singapore: Pan Stanford Publishing Pte. Ltd. http://doi.org/10.1201/b19751-14.

Smith, A. M. & Senior, J. J. (2021). Alginate Hydrogels with Tuneable Properties. In *Tunable Hydrogels. Advances in Biochemical Engineering Biotechnology*, edited by Antonina Lavrentieva, Iliyana Pepelanova and Dror Seliktar. Cham: Springer. https://doi.org/10.1007/10_2020_161.

Sneha Nair, M., Tomar, M., Punia, S., Kukula-Koch, W., Kumar, M. (2020). Enhancing the functionality of chitosan- and alginate-based active edible coatings/films for the preservation of fruits and vegetables: A review. *International Journal of Biological Macromolecules*, *164*, 304-320. https://doi.org/10.1016/j.ijbiomac.2020.07.083.

Stoica, R., Pop, S. F. & Ion, R. M. (2013). Evaluation of natural polyphenols entrapped in calcium alginate beads prepared by the ionotropic gelation method. *Journal of Optoelectronics and Advanced Materials*, *15*, 893-898.

Sun, J. & Tan. H. (2013). Alginate-Based Biomaterials for Regenerative Medicine Applications. *Materials*, *6*, 1285-1309. http://doi.org/10.3390/ma6041285.

Sun-Waterhouse, D., Penin-Peyta, L, Wadhwa, S. S. & Waterhouse, G. I. N. (2012). Storage Stability of Phenolic-Fortified Avocado Oil Encapsulated Using Different Polypmer Formulations and Co-extrusion Technoloy. *Food Bioprocessing and Technology*, *5*, 3090-3102. http://doi.org/10.1007/s11947-011-0591-x.

Szekalska, M., Puciłowska, A., Szymańska, E., Ciosek, P. & Winnicka, K. (2016). Alginate: current use and future perspectives in pharmaceutical and biomedical applications. *International Journal of Polymer Science.* http://dx.doi.org/10.1155/2016/7697031.

Tam, S. K., Dusseault, J., Bilodeau, S., Langlois, G., Halle, J. P. & Yahia, L. (2011). Factors influencing alginate gel biocompatibility. *Journal of Biomedical Materials Research Part A*, *98*, 40-52. http://doi.org/10.1002/jbm.a.33047.

Tønnesen H. H. & Karlsen, J. (2002). Alginate in Drug Delivery Systems. *Drug Development and Industrial Pharmacy*, *28*, 621-630. http://doi.org/10.1081/ddc-120003853.

Tzatsi, P. & Goula, A. M. Encapsulation of Extract from Unused Chokeberries by Spray Drying, Co-crystallization, and Ionic Gelation. *Waste and Biomass Valorization*, *12*, 4567-4585. https://doi.org/10.1007/s12649-020-01316-7.

Whelehan, M. & Marison, I. W. (2011). Microencapsulation using vibrating technology. *Journal of Microencapsulation*, *28*, 669-688. http://doi.org/10.3109/02652048.2011.586068.

Yun, P., Devahastin, S. & Chiewchan, N. (2021). Microstructures of encapsulates and their relations withencapsulation efficiency and controlled release of bioactiveconstituents: A review. *Comprehensive Reviews in Food Science and Food Safety*, *20*, 1768–1799. https://doi.org/10.1111/1541-4337.12701.

Zam, W., Bashour, G., Abdelwahed, W. & Khayata, W. (2014). Alginate-pomegranate peels' polyphenols beads: effects of formulation parameters on loading efficiency. *Brazilian Journal of Pharmaceutical Sicences*, *50*. http://dx.doi.org/10.1590/S1984-82502014000400009.

Biographical Sketch

Ina Ćorković

Affiliation: University of Josip Juraj Strossmayer in Osijek, Faculty of Food Technology Osijek, Osijek, Croatia.

Education: Faculty of Food Technology, Josip Juraj Strossmayer University – Master's degree.

Business Address: F. Kuhača 18, 31000 Osijek, Croatia.

Research and Professional Experience:
Project Research Assistant:
- Design, fabrication and testing of biopolymer gels as delivery systems for bioactive and volatile compounds in innovative functional foods; project financed by Croatian Science Foundation (January 2020 – December 2023).

Professional Appointments: Research Assistant at Faculty of Food Technology: 2020 –

Publications from the Last 3 Years:
1. Šafranko, S., Ćorković, I., Jerković, I., Jakovljević, M., Aladić, K., Šubarić, D., & Jokić, S. (2021). Green Extraction Techniques for Obtaining Bioactive Compounds from Mandarin Peel (Citrus unshiu var. Kuno): Phytochemical Analysis and Process Optimization. *Foods*, *10*(5), 1043.
2. Ćorković, I., Pichler, A., Šimunović, J., & Kopjar, M. (2021). Hydrogels: Characteristics and Application as Delivery Systems of Phenolic and Aroma Compounds. *Foods*, *10*(6), 1252.
3. Kopjar, M., Ivić, I., Buljeta, I., Ćorković, I., Vukoja, J., Šimunović, J., & Pichler, A. (2021). Volatiles and Antioxidant Activity of Citrus Fiber/Blackberry Gels: Influence of Sucrose and Trehalose. *Plants*, *10*(8), 1640.
4. Ćorković, I., A. Pichler, I. Buljeta, J. Šimunović, M. Kopjar: Carboxymethylcellulose hydrogels: Effect of its different amount on preservation of tart cherry anthocyanins and polyphenols. *Current Plant Biology* 28, 100222 (2021).
5. Kopjar, M., I. Buljeta, I. Ćorković, V. Kelemen, J. Šimunović, A. Pichler: Plant-based proteins as encapsulating materials for glucosyl-hesperidin. *International Journal of Food Science and Technology* 57, 728-737 (2022).
6. Ćorković, I., A. Pichler, J. Šimunović, M. Kopjar: Microencapsulation of glucosyl-hesperidin in alginate/chitosan hydrogel beads. *Food in Health and Disease* 2021, 10 (2), 39-44.

Mirela Kopjar, PhD

Affiliation: Faculty of Food Technology, Josip Juraj Strossmayer University, F. Kuhača 18, 31000 Osijek, Croatia.

Education: PhD - Faculty of Food Technology, Josip Juraj Strossmayer University.

Research and Professional Experience:
Principle investigator:
- Fibers and proteins as building blocks for development of novel bioactive food ingredients; project financed by European Social Fund and Croatian Science Foundation (October 2019 – April 2023)
- Design, fabrication and testing of biopolymer gels as delivery systems for bioactive and volatile compounds in innovative functional foods; project financed by Croatian Science Foundation (January 2020 – December 2023)
- Trehalose: fruit product quality improvement; project financed by Croatian Science Foundation (September 2014 – August 2017).

Professional Appointments:
- Full professor at Faculty of Food Technology: 2016 -
- Associate professor at Faculty of Food Technology: 2012 – 2016
- Assistant professor at Faculty of Food Technology: 2008 – 2012
- Assistant at Faculty of Food Technology: 2002 – 2008.

Publications from the Last 3 Years:
1. Kopjar, M., J. Šimunović, I. Ivić, J. Vukoja, A. Pichler: Retention of linalool and eugenol in hydrogels. *International Journal of Food Science and Technology*, 55, 1416-1425 (2020).
2. Vukoja, J., A. Pichler, I. Ivić, J. Šimunović, M. Kopjar: Cellulose as a delivery system of raspberry juice volatiles and their stability. *Molecules*, 55, 1416-1425 (2020).
3. Barišić, V., I. Flanjak, M. Kopjar, M. Benšić, A. Jozinović, J. Babić, D. Šubarić, B. Miličević, K. Doko, M. Jašić, Đ. Ačkar: Does high voltage electrical discharge treatment induce changes in tannin and fiber properties of cocoa shell? *Foods*, 9, 810 (2020).

4. Vukoja, J., I. Buljeta, A. Pichler, J. Šimunović, M. Kopjar: Formulation and stability of cellulose-based delivery systems of raspberry phenolics. *Processes*, 9, 90 (2021).
5. Ivić, M. Kopjar, Jakobek L., V. Jukić, S. Korbar, B. Marić, J. Mesić, A. Pichler: Influence of processing parameters on phenolic compounds and color of Cabernet Sauvignon red wine concentrates obtained by reverse osmosis and nanofiltration. *Processes*, 9, 89 (2021).
6. Vukoja, J., I. Buljeta, I. Ivić, A. Pichler, J. Šimunović, M. Kopjar: Disaccharide Type Affected Phenolic and Volatile Compounds of Citrus Fiber-Blackberry Cream Fillings. *Foods*, 10(2), 243, (2021).
7. Ivić, I., M. Kopjar, V. Jukić, M. Bošnjak, M. Maglica, J. Mesić, A. Pichler: Aroma Profile and Chemical Composition of Reverse Osmosis and Nanofiltration Concentrates of Red Wine Cabernet Sauvignon. *Molecules*, 26, 874 (2021).
8. Grgić, I., M. Grec, A. Gryszkin, T. Zięba, M. Kopjar, Đ. Ačkar, A. Jozinović, B. Miličević, S. Zavadlav, J. Babić: Starches modified by combination of phosphorylation and high-voltage electrical discharge (HVED) treatment. *Polish Journal of Food and Nutrition Science*, 71(1), 79–88 (2021).
9. Ivić, I., M. Kopjar, J. Obhođaš, A. Vinković, D. Pichler, J. Mesić, A. Pichler: Concentration with nanofiltration of red wine Cabernet Sauvignon produced from conventionally and ecologically grown grapes: Effect on volatile compounds and chemical composition. *Membranes*, 11(5), 320 (2021).
10. Ivić, I., M. Kopjar, D. Pichler, I. Buljeta, A. Pichler: Concentration with nanofiltration of red wine Cabernet Sauvignon produced from conventionally and ecologically grown grapes: Effect on phenolic compounds and antioxidant activity. *Membranes*, 11(5), 322 (2021).
11. Ćorković, I., Pichler, A., Šimunović, J., & Kopjar, M. Hydrogels: Characteristics and Application as Delivery Systems of Phenolic and Aroma Compounds. *Foods*, *10*(6), 1252 (2021).
12. Buljeta, I., Pichler, A., Šimunović, J., & Kopjar, M. Polyphenols and Antioxidant Activity of Citrus Fiber/Blackberry Juice Complexes. *Molecules*, 26(15), 4400 (2021).
13. Buljeta, I., Pichler, A., Ivić, I., Šimunović, J., & Kopjar, M. Encapsulation of Fruit Flavor Compounds through Interaction with Polysaccharides. *Molecules*, 26(14), 4207 (2021).
14. Kopjar, M., Ivić, I., Buljeta, I., Ćorković, I., Vukoja, J., Šimunović, J., & Pichler, A. Volatiles and Antioxidant Activity of Citrus

Fiber/Blackberry Gels: Influence of Sucrose and Trehalose. *Plants*, 10(8), 1640 (2021).
15. Buljeta, I., A. Pichler, J. Šimunović, M. Kopjar: Polyphenols and antioxidant activity of citrus fiber/blackberry juice complexes. *Molecules*, 26(15), 4400 (2021).
16. Ćorković, I., A. Pichler, I. Buljeta, J. Šimunović, M. Kopjar: Carboxymethylcellulose hydrogels: Effect of its different amount on preservation of tart cherry anthocyanins and polyphenols. *Current Plant Biology* 28, 100222 (2021).
17. Kopjar, M., I. Buljeta, I. Jelić, V. Kelemen, J. Šimunović, A. Pichler: Encapsulation of cinnamic acid on plant-based proteins: Evaluation by HPLC, DSC and FTIR-ATR. *Plants* 10, 2158 (2021).
18. Kelemen, V., A. Pichler, I. Ivić, I. Buljeta, J. Šimunović, M. Kopjar: Brown rice proteins as delivery system of phenolic and volatile compounds of raspberry juice. *International Journal of Food Science and Technology*. 57, 1866–1874 (2022).
19. Kopjar, M., I. Buljeta, I. Ćorković, V. Kelemen, J. Šimunović, A. Pichler: Plant-based proteins as encapsulating materials for glucosyl-hesperidin. *International Journal of Food Science and Technology* 57, 728-737 (2022).
20. Ćorković, I., A. Pichler, J. Šimunović, M. Kopjar: Microencapsulation of glucosyl-hesperidin in alginate/chitosan hydrogel beads. *Food in Health and Disease* 2021, 10 (2), 39-44.

Chapter 5

Alginate as a Carrier for Probiotic Immobilization: Extrusion and Spray Dry Technique

Tanja Krunic[*]
Innovation Centre Faculty of Technology and Metallurgy,
University of Belgrade, Belgrade, Serbia

Abstract

Alginate is a natural polysaccharide, which can form a gel and has simple preparation conditions and processes. Also, alginate is a widely used carrier in the food industry due to its pH responsivity, which is important for gastrointestinal digestion. It has the characteristics of biosafety, and biocompatibility and has been widely used and studied in recent years. Alginate is used for probiotic immobilization by extrusion technique due to its ability to form a gel in mild conditions with satisfied the mechanical and chemical stability of carriers. The paper aims to summarize different uses of alginate as a carrier by extrusion and spray dry technique. The viability of probiotic cells is very important due to their numerous health benefits. The desired number of viable bacteria is difficult to achieve as their number decreases due to many environmental factors. Probiotic immobilization increases the viability of bacteria against unfavorable environmental conditions during processing, storage, and gastric and intestinal digestion (conditions such as low pH, presence of digestive enzymes, and bile salt in a human gastrointestinal tract). Different immobilization techniques and the addition of many agents to alginate carriers create beads in a wide range of diameters, with changed fermentative activity and viability of probiotics. This paper

---

[*] Corresponding Author's Email: tkrunic@tmf.bg.ac.rs.

In: Properties and Applications of Alginate
Editor: Michael Y. Wilkerson
ISBN: 979-8-88697-371-6
© 2022 Nova Science Publishers, Inc.

reviews the application of sodium alginate in the fields of the probiotic food industry.

Keywords: probiotic, immobilization, alginate, extrusion, spray dry

Introduction

In the recent decade, there has been an explosion of probiotic health based products. The probiotics market was valued at USD 48.88 billion in 2019 and is estimated to increase to USD 94.48 billion by 2027. This expansion is owing to consumer perception about the health benefits of probiotic-based products, in addition to the demand for immunity-boosting products amid COVID-19 (Market Research Report, 2020). Probiotics are microorganisms considered to be generally recognized as safe (GRAS) for human use and include a large range of bacteria and some yeasts (strains of genera *Lactobacillus, Bifidobacterium, Enterococcus, Streptococcus, Lactococcus*, and yeasts *Saccharomyces cerevisiae, Saccharomyces boulardii*) (Krunic 2021). Mostly used probiotics in beverages such as yogurt are *Lactobacillus acidophilus, Lactobacillus rhamnosus, Lactobacillus delbrueckii* ssp. *Bulgaricus* and *Bifidobacterium bifidum*. The shelf lives of probiotics should be controlled in order to manufacture products with a satisfactory number of live bacteria (at least 10^6 to 10^7 CFU/g) to obtain the health benefits of probiotic cultures. Unfortunately, many studies indicated that there is poor survival of probiotic bacteria in these products. Most probiotic food products are categorized as functional foods because probiotics have many impacts on human health and the body. They have immunomodulatory properties that usually act directly by modulating the secretion of immunoglobulins or cytokines, increasing the activity of macrophages or natural killer cells, or indirectly by enhancing the gut epithelial barrier, altering the mucus secretion, and competitive exclusion of other bacteria (Krunic, 2021). Probiotics in general have a low survival rate and little resistance to different processing techniques and environmental and gastrointestinal conditions, such as temperature, pH, and bile salt. According to the definition, to be beneficial functional foods, the probiotic bacteria should survive passage through the stomach and reach their site of action in adequate amounts. In the dairy industry, immobilization has been applied to improve the survival and delivery of bacterial cultures (Krunic et al., 2016; Krunic et al., 2017; Krunic et al., 2019).

Immobilization is a valuable method that has been recognized for use in the food industry. Immobilization of probiotics brings many benefits such as probiotics in a carrier are protected from external factors, so they have a greater ability to survive manufacture and storage time and conditions. Immobilization increases the viability of probiotic bacteria against unfavorable environmental conditions during processing and storage (Krunic et al., 2016). Also, immobilized probiotics can survive conditions such as low pH, the presence of digestive enzymes, and bile salt in a human gastrointestinal tract (Krunic, et al., 2016; Krunic, et al., 2019). The technology of immobilization of probiotics meets an important requirement: bacterial cells must stay alive. This requirement has been crucial in selecting the appropriate immobilization technology. Emulsification and extrusion are technologies of encapsulation that best satisfy the stated condition. Emulsification allows the production of a wide particle size range from 0.2 to 5000 μm whereas, extrusion gives a smaller range size and provides more uniform beads (Burgain, et al., 2011; Krunic et al., 2019).

The disadvantage of the extrusion technique is difficult to use in large-scale productions due to the slow formation of the carriers with encapsulated cells. Spray drying is one of the most used techniques in the microencapsulation process, due to its short time, low cost, good stability and quality of end products, and applicability to large-scale production. This technique enables rapid cell encapsulation in industrial conditions, but the disadvantage of this method is the influence of hot dry air on the cell viability, which is why optimization of process parameters is necessary (Barbosa et al., 2015).

Alginate as Carrier

Alginate is the most widely used encapsulating material. Alginate is a linear polysaccharide consisting of β-Dmannuronic and α-L-guluronic acids. Alginates with a wide range of compositions can be obtained from bacteria such as *Azotobacter vinelandii* and *Pseudomonas* spp., or commercially from brown marine macroalgae (Kjesbu et al., 2022). The alginate M/G ratio, total biopolymer concentration, cation type, and concentrations have a great influence on the textural and mechanical properties of the encapsulating material. Encapsulation of living cells is limited to carriers that can build nets under mild conditions which is alginate's main characteristic. Sodium alginate may be ionically crosslinked with multivalent cations to form gels. Alginate beads prepared from sodium alginate of high guluronic acid content were found to have greater stiffness since most of the divalent cations bind to the guluronic sequences but only to an extent or not at all to the mannuronic acid and alternating blocks. Increasing the concentration of alginate and gelling ions also generated a similar effect (Chana et al., 2011). Figure 1 shows MM, MG, and GG blocks in the alginate molecule.

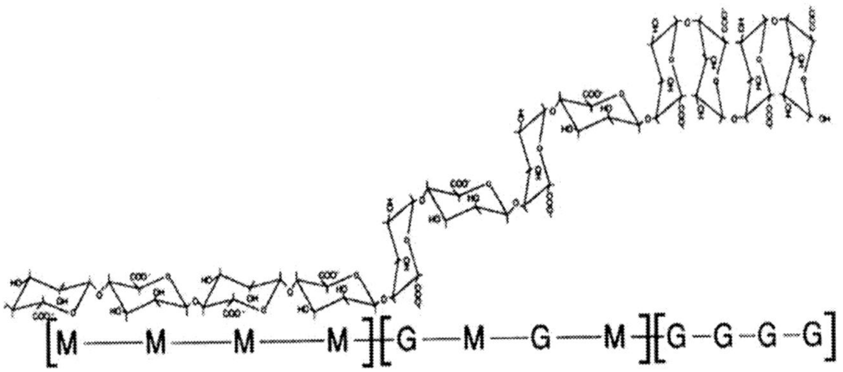

Figure 1. Alginate molecule.

Alginate hydrogel beads can be produced by the extrusion of alginate and probiotic mixture through a needle with an electrostatic potential between the needle and crosslinking solution (usually used $CaCl_2$) (Krunic et al., 2016; Krunic et al., 2019).

Inherent advantages of an alginate (and many other hydrogels) for probiotic encapsulation can be listed (Krunic at al., 2019; Pérez-Luna and González-Reynoso, 2018).

- the aqueous environment helps maintain the biological function of the encapsulated material;
- the crosslinked nature of hydrogel provides a diffusion barrier capable of allowing the passage of molecules of a given size threshold while excluding larger molecules from interacting with the encapsulated probiotic cells;
- the size threshold can be tailored according to the degree of crosslinking of the hydrogel;
- hydrogels can be created in situ under mild reaction conditions (temperature and pH), which make them convenient for minimally invasive surgery;

- alginate carrier can be created to meet the needs of different carrier properties according to specific applications and can be mixed with many other molecules in order to provide the required carrier;
- alginate can provide an adequate 3-D microstructure that can modulate the mechano-biochemical transduction signals of encapsulated cells.

As previously indicated the mechanical and chemical stability of alginate beads could be improved by using different materials for blending with or coating alginate carriers. The addition of different materials could enable greater control over the bacterial release and improve the viability of encapsulated probiotics (Krunic et al., 2016; Krunic, 2017; Krunic et al., 2019). Chitosan, the cationic (1-4)-2-amino-2-deoxy-β-D-glucan, is industrially produced from marine and fungal chitin and is being used in the food industry. Alginate and chitosan are biodegradable, biocompatible, and nontoxic.

Chitosan has a high modulus along with low elongation-at-break but mixing or copolymerizing chitosan with different polymers can influence its morphology and plasticity (Brychcy et al., 2015).

In recent studies, the textural properties of biopolymer beads of different compositions were tested. These results indicated that differences in mechanical responses to deformation exist because of different crosslinking reactions between biopolymers networks. The textural and physical properties, as well as the entrapment efficiency of the carriers, were greatly affected by the total biopolymer concentration and the employed ratio between biopolymers (Krunic et al., 2016; Krunic et al., 2019; Krunic and Rakin 2021).

Whey protein is a less used material in extrusion technique for encapsulation but is mostly used in emulsion encapsulation or spray dry technique. Whey protein is a mixture of globular proteins isolated from whey as a by-product in cheese production. The ability of whey proteins to form gels without the use of heat treatment and any chemicals makes them an effective carrier for the protection of probiotics used in dairy and functional food (Krunic et al., 2018).

Encapsulation by Extrusion

Probiotics can be encapsulated using a variety of methods including physical methods (such as spray drying), and physicochemical methods (such as ionic gelation). An extrusion technique with sodium alginate is a physicochemical method of ionic gelation.

Encapsulation is a process to entrap one product or substance within another, thereby producing solid particles with different diameters (nanometers to millimeters). The product or substance that is encapsulated in the carrier (usually hydrogel). The material used for encapsulation is called the matrix, coating, or carrier material. In food products, all materials used in encapsulation processes should be food grade.

The benefits of the encapsulation technique applied in the food industry are many. For example improved stability in the final food product and during processing (no degradation or reaction with other components in the food and/or less evaporation of volatile active agent), off-taste masking, and controlled release (differentiation, release by the right stimulus). It is important that benefits overcome possible negative effects, such as additional costs, increased complexity of production process and/or supply chain, and undesirable consumer notice (visual or touch) of the encapsulates in final food products (Zuidam and Shimoni, 2010).

A number of efforts have been made to improve the viability of probiotics (and their functional effect) in an important variety of foods. Among these attempts, encapsulation is an emerging technology that could improve the functionality of these living microorganisms to be incorporated into different food matrices.

One of the problems of probiotics encapsulation technology is the bacterial size (typically 1–5 μm diameter), which immediately excludes nanotechnologies. Also, capsules over 1 mm are considered large and can produce a coarseness of texture in live microbial feed supplements, and the sizes less than 100 μm do not significantly protect the probiotics in adverse conditions (Petzold et al., 2017).

The size of capsules has been reported to be associated with encapsulation techniques.

Emulsification allows a smaller size of probiotic capsules to be obtained with a range of 0.1–8000 μm, for spray coating a range of 5–5000 μm, and for spray drying a narrow range of approximately 10–120 μm. The biggest capsule would be obtained by co-extrusion with a range of 120–9000 μm (Burgain et al., 2011).

The Extrusion Technique for Probiotic Immobilization

Extrusion is one of the most commonly used technology for probiotic encapsulation. The reason for that is the mild conditions required for carrier obtaining, low cost of technology (because do not require high temperature), and widely available biopolymers used as carriers. Hydrogels as alginate are ideal for the encapsulation of living cells by extrusion technique. What properties alginate beads have depends on hydropolymer concentration and beads diameter. Also, incorporating different molecules in carrier composition leads to a wide range of properties.

Figure 2 shows the encapsulation of probiotics by the extrusion technique.

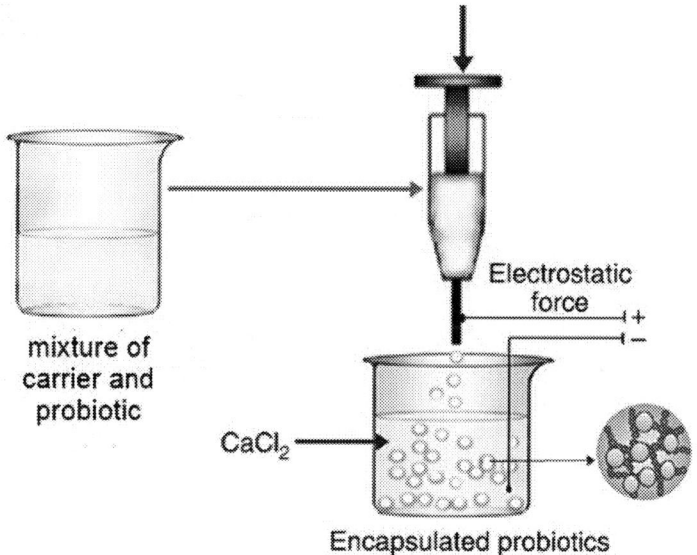

Figure 2. Electrostatic extrusion technique used for probiotic encapsulation by alginate based carrier.

Krunic et al., (2016) compared the fermentative activity of ABY 6 probiotic starter culture (*Streptococcus salivarius* ssp. *thermophilus* (80%), *Lactobacillus acidophilus* (13%), *Bifidobacterium bifidum* (6%), *Lactobacillus delbrueckii* ssp. *bulgaricus* (1%)) encapsulated in alginate and alginate beads coated with chitosan. Encapsulated ABY 6 probiotic culture showed lower fermentative activity which is expected because of matter diffusion through the carrier (Krunic et al., 2016; Krunic et al., 2017). Alginate beads coated with chitosan showed lower fermentative activity but higher stability during 28 days of storage and viability during SGI condition compared to alginate non-coated beads with ABY 6 probiotic starter culture (Krunic et al., 2016).

Lopes et al., (2021) also examined alginate carrier, and alginate carrier coated with chitosan as probiotic encapsulate material. The microencapsulation of the probiotic culture *Lactobacillus acidophilus* La-05 resulted in higher probiotic survival (> 6 log CFU/mL in product and simulated gastrointestinal conditions), and improved technological (no moisture loss, lower proteolysis, and organic acid content), texture (lower gumminess and adhesiveness), and volatile (compounds with floral and fruity notes and lower "goat" aroma) properties. Also, chitosan coating did not improve the effects of encapsulated probiotic culture used for cheese production (Lopes et al., 2021). Figure 3 shows alginate and alginate coated beads.

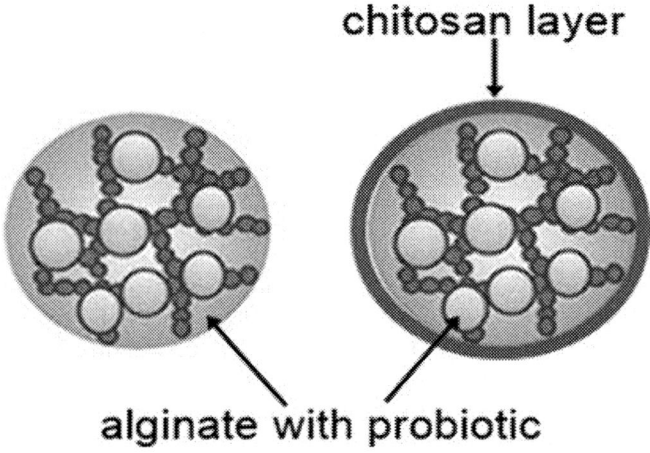

Figure 3. Alginate and alginate coated beads used for probiotic encapsulation.

A combination of alginate and whey protein as well as whey protein hydrolysate showed better fermentative activity, stability, and probiotic viability during SGI conditions compared to alginate and alginate beads coated with chitosan (Krunic et al., 2019; Krunic et al., 2016). Krunic et al. (2019) encapsulated ABY 6 probiotic starter culture by electrostatic extrusion using negatively charged alginate enriched with whey protein concentrate (WPC) and whey protein hydrolysate (WPH). Divalent cations used for alginate gelation, such as calcium, can induce aggregation of the protein. The protein or peptide aggregation occurs in three different ways: (1) electrostatic shielding; (2) ion/hydrophobic interactions; and (3) crosslinking with negatively charged carboxylic groups of whey protein molecules, via protein – cation – protein bridges (Ramos et al., 2014). Whey protein – alginate beads are presumed to be formed through a cross-linking process among proteins, alginate, and calcium ions. The addition of different biopolymers leads to a change in the carrier properties. Enriching the alginate matrix with whey protein and peptides leads to increase mechanical properties of the carrier, and coating alginate beads with chitosan increases mechanical properties even more (Krunic et al., 2016; Krunic et al., 2019). Although the mentioned enrichment of alginate carriers has a positive effect on the mechanical properties of the beads, when it comes to fermentative activity and carrier porosity, the additives have different effects. While chitosan reduces carrier porosity and slows down fermentation, the addition of WPC and WPH has a positive effect on fermentation but also increases the release of probiotics from the matrix due to higher carrier porosity (Krunic et al., 2016; Krunic et al., 2019).

Beads' diameter and structure are changing during the fermentation process due to the metabolic activity of encapsulated bacteria and the process depends on the carrier composition. The beads with carriers made of alginate and WPC changed diameter from 799 ± 25 μm to 1078 ± 28 μm and the diameter of beads with carriers made of alginate and WPH has changed from 745 ± 27 μm to 988 ± 26 μm (Krunic et al., 2019). This change in beads diameter is accompanied by an increase in the hardness of the beads (Table 1) as well as a change in the FT-IR fingerprint.

Table 1. Mechanical strength was shown via maximal force measured by compressions testing of packed beads
(Krunic et al., 2016; Krunic, 2017; Krunic et al., 2019)

	Maximal force, (30% deformation) before fermentation, N	Maximal force, (30% deformation) after fermentation N	Cell release during fermentation, %
Alginate	2,1741 ± 0,015	2,3841 ± 0,024	3,44 ± 0,19
Alginate + chitosan	2,8731 ± 0,028	2,9632 ± 0,027	2,49 ± 0,18
Alginate+WPC	2,3861 ± 0,041	2,9541 ± 0,057	19,70 ± 0,78
Alginate+WPH	2,3591 ± 0.021	2,781 ± 0,030	6,98 ± 0,23

Table 1 shows the results of the mechanical test obtained at a compression speed of 0.25 mms^{-1} min and mentioned sample deformation for alginate beads with immobilized probiotic cells showed that the measured maximal forces increased after fermentation. During the fermentation, the probiotic culture produces lactic acid, which decreases the pH of the medium, which leads to the creation of conditions that adversely affect the mechanical strength of the beads. Krunic et al., (2016) explained that the mechanical strength of particles increased because alginate builds bonds with calcium ions from milk and whey which have a positive impact on beads' strength. That is one reason for increased mechanical properties, but also what is not mentioned is that decreasing pH has a strong impact on carboxyl groups of alginate, which leads to a contracted hydrogel because the polymer chains come close together but in the research described by Krunic et al., (2019), it was found that beads do not shrink during fermentation, but rather increase in diameter. Therefore, it can be concluded that an increase in mechanical strength of beads during fermentation is a consequence of re-bonding between protein and alginate, as a result of the metabolic and proteolytic activity of encapsulated culture. This is confirmed by FTIR analysis (Krunic et al., 2019; Krunic and Rakin 2021). Comparing the results of the mechanical test from Table 1 it is clear that the addition of WPC and WPH in alginate contributes to the improvement of mechanical properties of beads before fermentation and leads to a greater increase in mechanical strength during fermentation.

The increased fermentative activity of the encapsulated culture was accompanied by the increased release of cells from the matrix (Krunic et al., 2016; Krunic, 2017; Krunic et al., 2019). The high porosity of protein – alginate beads is also noticeable through cell release. The alginate-WPC matrix showed more cell release (18.10 ± 0.78%) compared to Aalginate – WPH matrix (11.66 ± 0.23%), indicating higher porosity of the first matrix. Also, both samples with protein and peptides showed higher cell release than the alginate and alginate – chitosan matrix (Krunic et al., 2016; Krunic, 2017; Krunic et al., 2019). Greater cell release for alginate – WPC matrix compared to alginate – WPH indicates that native whey protein with alginate provides a very porous matrix, while shorter peptide chains provide a less porous carrier. The porosity of the carrier does not present a disturbance in fermentation, because despite the high percentage of cell release from the alginate – WPC sample the number of cells in the beads is in the same range as in the alginate – WPH carrier (with cell release 11.66 ± 0.23%) and higher than a number of viable encapsulated cells in the case of alginate or alginate – chitosan carrier (Krunic et al., 2016; Krunic, 2017; Krunic et al, 2019).

Previously, it was concluded that the presence of proteins and peptides in an alginate matrix used for probiotic encapsulation contributes to mechanical properties of carrier, fermentative activity, better acid, and bile tolerance, as well as better survival of encapsulated probiotics during the simulated gastrointestinal condition. As proteins are the main source of bioactive peptides, the continuation of the previously mentioned research (Krunic et al., 2019) was based on determining the antioxidant capacity of the carrier used for probiotic encapsulation (Krunic and Rakin, 2022).

The addition of whey protein and whey protein hydrolysate in the alginate matrix was not significantly affected the efficiency of encapsulation, but alginate – WPH showed slightly better efficiency of encapsulation compared to alginate – WPH (Krunic and Rakin, 2022).

The fermentation of the whey based substrate with encapsulated ABY 6 culture results in increased antioxidant capacity of the beverage.

The effect is not the same in samples with different carriers that were used for probiotic encapsulation. Enriching an alginate carrier with the bioactive component is an excellent way to increase the beverage's antioxidant capacity. Encapsulated with alginate-WPC and alginate-WPH improve antioxidant and probiotic stability of beverage during 28 days of storage described by Krunic and Rakin (2022). The highest value of antioxidant capacity has the whey based substrate fermented with probiotic starter culture encapsulated in the carrier with hydrolyzed whey protein compared to alginate, alginate – chitosan, and alginate – WPC carrier (Krunic et al., 2017; Krunic and Rakin 2022).

Mirmazloum et al., (2021) co-encapsulate probiotic *L. acidophilus* La-14 cells with *Ganoderma lingzhi* extract to prolong the viability of the cells under simulated gastrointestinal (SGI) conditions and to protect the active ingredients of *Reishi* mushroom during the storage period. Combinations of sodium alginate, chitosan, maltose, hydroxyethylcellulose (HEC), hydroxypropyl methylcellulose (HPMC), and calcium lactate were tested as carriers' compounds. The significantly positive effect of maltose supplementation and the double-layer coating on reducing the release rate of the active molecules (phenolic compounds, antioxidants, and β-glucan) was confirmed. Significant improvement in probiotic cell viability under simulated gastrointestinal conditions has been found and confirmed by confocal laser microscopy in maltose containing double-layer coated calcium alginate beads variants (Mirmazloum et al., 2021).

Kjesbu et al., (2022) produced alginate-based beads by electrostatic extrusion technique with a diameter range between 0.4 and 0.7 mm and with different spherical properties depending on carrier composition. They showed that increasing the content of cellulose nanofibrils in the polymer mixture resulted in elongation during extrusion and generated microbeads with increased size, size dispersity, and higher aspect ratios. Ionic crosslinking using calcium alone resulted in beads with increasing content of cellulose nanofibrils exhibiting reduced stability. Compression of 1.8% (w/v) gel cylinders revealed that Young's modulus decreased when adding cellulose nanofibrils into alginate, but syneresis was reduced. Spectrophotometry using FITC dextrans revealed that initial uptake and release rates were slightly higher in microbeads with cellulose nanofibrils compared to alginate alone, indicating a slightly higher porosity. The study showed that composite alginate and tunicate cellulose nanofibrils microbeads can be produced with an electrostatic bead generator (Kjesbu et al., 2022). Such beads can be used for the encapsulation of a wide range of different types of cells.

Feng et al., (2020) showed double-layered encapsulation alginate based carrier fabricated to protect probiotics. *Lactobacillus plantarum* was successfully encapsulated in the fibers (10^7 CFU/mg) and exhibited an oriented distribution along the fiber. The encapsulation of core-shell fiber mat enhanced the tolerance of probiotic cells to simulated gastrointestinal conditions with no significant loss of viability. Also, the encapsulated probiotic cells exhibited better thermal stability under heat moisture treatment, and lower loss of viability (0.32 log CFU/mL) occurred when compared with the free or encapsulated probiotic cells in uniaxial fiber mat. Feng et al., (2020) concluded, that double-layered carrier presents great potential in probiotic encapsulation and improving their resistant ability to the harsh conditions.

Liu et al., (2020) examined the combination of fish gelatin and sodium alginate as a carrier for probiotic encapsulation. The results indicated that the addition of FG could improve the transparency, rehydration, and water-holding capacity of the alginate based carrier. Also, encapsulation efficiency increased from 15% to 92%. *Bifidobacterium longum* embedded by a fish gelatin/sodium alginate double network gelation showed better thermal stability than when it was free. Compared with bare probiotics (1.7%), the encapsulated ones exhibited higher viability (above 15%) in simulated gastric fluid (Liu et al., 2020).

The Spray Dry Technique for Probiotic Immobilization

Alginate based formulations have shown desirable functional characteristics for probiotic encapsulation. The main disadvantage of the extrusion method is that it does not produce dry long-lasting encapsulated probiotics. The spray drying technique solves that problem by obtaining dry, shelf-stable products. Spray drying is a physical method of probiotic encapsulation. Alginate as a carrier is not widely used in spray dry as much as in the extrusion technique. The reason for this is that alginate's main characteristic to form a gel under mild conditions is not important for this technique, except in some hybrid and modified techniques as one described by Tan et al., (2022).

Spray drying is a unit operation by which a liquid product is atomized in a hot gas current to instantaneously obtain a powder. The gas generally used is air or more rarely an inert gas. The initial liquid feeding the sprayer can be a solution, an emulsion, or a suspension (Gharsallaoui et al., 2007).

The spray drying scheme is shown in Figure 4.

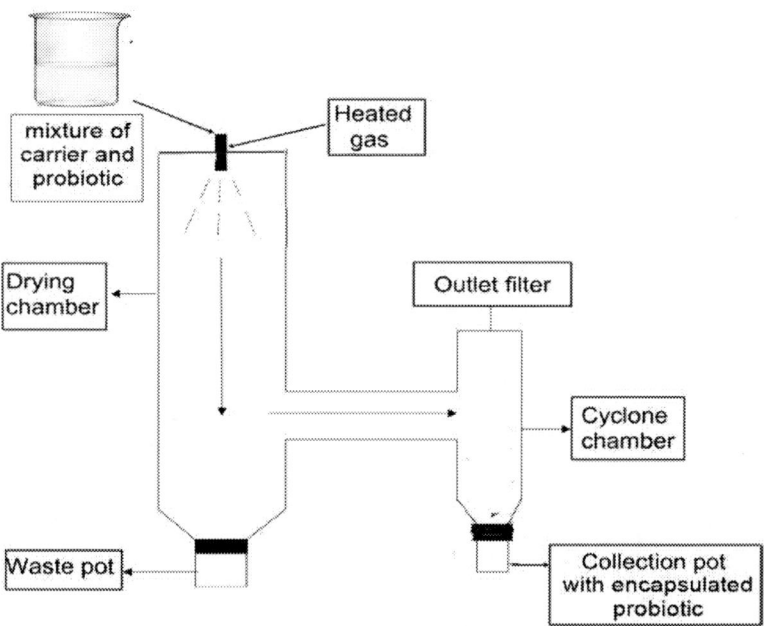

Figure 4. Spray dry technique used for probiotic immobilization by an alginate based carrier.

Probiotic powders obtained by spray drying are usually microbeads in the amorphous state (Wang et al., 2020). This amorphous structure is associated with the presence of carbohydrates in the dehydration media and has also been related to the improved stability of spray dried microorganisms (Vivek et al., 2020).

Tan et al., (2022) developed a spray drying technique that combines particle formation, alginate crosslinking, and drying into a single step, thereby streamlining the production of encapsulated probiotics powder. *Lacticaseibacillus rhamnosus* GG (LGG) was encapsulated in six encapsulation formulations (Sucrose, Alginate, Alginate – sucrose, Calcium – alginate, Calcium – alginate – sucrose, and Calcium – alginate – sucrose, using inverted feed channels) were characterized and compared by Tan et al., (2022). The crosslinked alginate with sucrose formulation (CAS) was found to be the most promising, achieving ~10^9 CFU/g of surviving *L. rhamnosus* GG after spray drying and exposure to simulated gastric fluid (SGF). Similar results were obtained with *Lactiplantibacillus plantarum* and *Lacticaseibacillus paracasei,* encapsulated in CAS carrier (Tan et. al., 2022).

Yonekura et al., (2014) evaluated sodium alginate, chitosan, and hydroxypropyl methylcellulose (HPMC) as carrier components for spray dried *Lactobacillus acidophilus* NCIMB 701748 by assessing their impact on cell viability and physicochemical properties of the dried powders, viability over 35 days of storage at 25 °C and survival after simulated digestion. Previously mentioned materials were added to a control carrier medium containing whey protein concentrate, d-glucose, and maltodextrin. Alginate and HPMC maintained *L. acidophilus* viability during spray drying but decreased survival during simulated digestion.

Hadzieva et al., (2017) have developed microcapsules based on soy protein isolate and alginate for microencapsulation of *Lactobacillus casei* using the spray drying method. Encapsulated probiotics showed the viability of the probiotic of 11.67 log CFU/g after microencapsulation, 10.05 log CFU/g after 3 h in simulated gastric conditions, 9.47 log CFU/g after 3 h in simulated intestinal conditions, and 9.20 log CFU/g after 4-month cold storage.

Chang et al., (2020) explored the possibility of improving the viability of *Akkermansia muciniphila* by encapsulating it in succinate-grafted alginate doped with epigallocatechin-3-gallate (EGCG). In that study, the determined surface properties of microcapsules and modified materials and the measured viability of probiotics after spray drying showed that the modified sodium alginate made the surfaces of microcapsules smoother and denser during the spray drying, thus preventing damage. EGCG enhanced the antioxidant capacity of probiotics by filling the pores inside microgels. They analyzed the long-term storage vitality changes, oxidation resistance, uniformity, particle size, and Zeta potential of microcapsules and found that spray dried modified sodium alginate microcapsules with EGCG showed better storability and stability. Also, they showed that EGCG-modified sodium alginate microbeads improved the protection of probiotic cells from the gastrointestinal fluid.

In the study described by Rajam et al., (2012) two different wall materials (whey protein isolate with sodium alginate and denatured whey protein isolate with sodium alginate) were used for microencapsulation of *Lactobacillus plantarum* (mtcc 5422) using spray drying and freeze-drying techniques. It was observed that the moisture content of the spray dried powder was lower than freeze-dried powder, but the spray drying method produced 9 – 12% less cell survival compared to the freeze-drying method. In simulated acidic and bile conditions carriers built by denatured whey protein and sodium alginate showed better protection of probiotic cells than carriers obtained by un-denatured whey protein and sodium alginate.

The objective of the research described by Liu et al., (2018) was to investigate the in vitro gastrointestinal digestion and storage properties of probiotic bacteria (*Lactobacillus zeae* LB1) encapsulated in microbeads containing sodium alginate (AG) and sodium caseinate (NaCas) or soy protein isolate (SPI) through spray drying technology. A high survival rate (>80%) was obtained for all the encapsulation formulas tested. A combination of SPI or NaCas with AG showed a positive effect on both the in vitro gastric resistance and storage stability of *L. zeae* LB1. In addition, SPI-AG microbeads provided the best protection to the probiotic cells during gastric digestion, whereas the best storage property was provided by NaCas-AG microbeads.

Obradovic (2019) used the spray dry technique with the same carrier (alginate and alginate with whey protein) and probiotic mixture (commercial probiotic starter culture ABY 6) as used in the papers of Krunic et al., (2016), Krunic et al., (2019), and Krunic and Rakin (2022). Comparing previously mentioned papers can be concluded that both techniques have advantages and disadvantages. The SEM micrographs showed that the carriers after spray drying retain a spherical shape with dents on the system surface. The surface of the carrier made of alginate and WPC was with smaller dents compared to the micrographs obtained for the alginate carriers without WPC. The reason for this is that the high protein content additionally affected crust formation. Also, a high drying rate leads to rapid wall solidification due to protein denaturation during the process. So, the presence of protein in the carrier influenced the formation of smother carrier surfaces (Obradovic, 2019). Encapsulation efficiency for spray dry and electrostatic extrusion techniques is shown in Table 2. If it is known that the product yield of the extrusion technique is almost 100%, based on the results shown in Table 2 it can be concluded that electrostatic extrusion is a better option for encapsulation of probiotic starter culture ABY 6.

Table 2. The efficiency of encapsulation end product yield of alginate and WPC-alginate beads obtained by electrostatic extrusion and spray dry technique (Krunic and Rakin, 2022; Obradovic, 2019)

Technique	Carrier	Encapsulation efficiency, %	Product yield, %
Electrostatic extrusion	Alginate	94.20 ± 0.61	/
	Alginate+WPC	92.98 ± 0.56	/
Spray dry	Alginate	76.8 ± 0.04	70.3 ± 0,2
	Alginate+WPC	79.6 ± 0.03	59.8 ± 0.3

In order to have a broader picture, it is necessary to compare the fermentable activity of the encapsulated culture, as well as the stability of the product during storage and gastrointestinal conditions.

Encapsulated probiotics by electrostatic extrusion technique showed slightly better fermentation activity and approximately the same stability during storage compared to probiotics encapsulated by spray dry technique. What is the main difference is the viability of encapsulated probiotics during the simulated gastrointestinal condition. Probiotics encapsulated by the electrostatic extrusion technique (Krunic et al., 2019) showed higher viability compared to one encapsulated by the spray dry technique (Obradovic, 2019).

The size of the carrier is important because as the size of the carrier decreases, the surface of the carrier that is exposed to moisture increases, which has a negative impact on the viability of the encapsulated culture. The carrier size obtained by spray dry techniques does not influence the sensory properties of the beverages.

Comparing alginate and alginate – WPC carriers, the alginate carriers with WPC showed lower values of zeta potential (Obradovic, 2019), which indicates their contribution to system chemical stability. The higher stability can be explained by establishing new bonds.

Conclusion

Alginate is a hydrogel that has a great ability to be used as a carrier for probiotic encapsulation by extrusion technique. A variation of alginate concentration and the addition of different molecules (proteins or other biopolymers) is an easy way to change carrier properties and adapt them to the required properties.

In the spray dry technique, alginate is not so much represented as a basic carrier as an adjunct to an encapsulation carrier that is predominantly protein in order to improve some characteristics. Milk and whey protein have been widely used in spray dry techniques, but in spray dry, as in the extrusion technique, a combination of protein and alginate is not uncommon.

Alginate as a carrier (either as a basic, dominant component or as an additive) has proven to be an excellent material for protecting probiotics during production, storage, and digestion. Immobilization of probiotics in an alginate based carrier significantly increased the probiotic properties of the product.

Protein-alginate delivery systems for encapsulation of probiotics and bioactive compounds have been a growing interest in recent years. Technological and functional properties of proteins, such as the ability of whey proteins to form gels without the use of severe heat treatment and toxic chemicals, make them an attractive material for controlled delivery applications in the food industry, as well as an effective carrier for the protection of probiotic used in functional food (Krunic, Rakin, Bulatovic, and Zaric, 2018). Using proteins, especially peptides, as a matrix component achieved three objectives: protection of probiotics, enrichment of products with antioxidants, and neutralization of possible bitter taste (because the bitter-tasting peptides are incorporated into the matrix and as such cannot contribute to the taste of the product) that bioactive peptides usually possess.

Acknowledgment

This work was supported by the Ministry of Education, Science and Technological Development of the Republic of Serbia (Contract No. 451-03-68/2022-14/200287.

References

Barbosa, J., Borges, S., Amorim, M., Pereira, M. J., Oliveira, A., Pintado, M.E., Teixeira, P. (2015). Comparison of spray drying, freeze drying and convective hot air drying for the production of a probiotic orange powder. *Journal of Functional Foods 17*, 340–351.

Brychcy, E., Kulig, D., Zimoch-Korzycka, A., Marycz, K., Jarmoluk, A. (2015). Physicochemical properties of edible chitosan/hydroxypropyl methylcellulose/lysozyme films incorporated with acidic electrolyzed water. *International Journal of Polymer Science*, 604759.

Burgain, C., Gaiani, C., Linder, M., Scher, J. (2011). Encapsulation of probiotic living cells: from laboratory scale to industrial applications. *Journal of Food Engendering 104*, 467–483.

Chan, E. S. Limb, T., Voob, W., Pogakub, R., Teyc, B., Zhangd, Z. (2011). Effect of formulation of alginate beads on their mechanical behavior and stiffness. *Particuology 9*, 228–234.

Chang, Y., Yang, Y., Xu, N., Mu, H., Zhang, H., and Duan, J. (2020). Improved viability of *Akkermansia muciniphila* by encapsulation in spray dried succinate-grafted alginate doped with epigallocatechin-3-gallate. *International Journal of Bioogycall Macromolecules* 373-382.

Feng, K., Huang, R. M., Wu, R. Q., Wei, Y. S., Zong, M. H., Linhardt, R. J., Wu, H. (2020). A novel route for double-layered encapsulation of probiotics with improved viability under adverse conditions. *Food Chemistry* 125977.

Gharsallaoui, A., Roudaut, G., Chambin, O., Voilley, A., and Saurel, R. (2007). Applications of spray-drying in microencapsulation of food ingredients: an overview. *Food Research International 40*, 1107-1121.

Hadzieva, J., Mladenovska, K., Crcarevska, M. S., Dodov, M. G., Dimchevska, S., Geskovski, N., et al. (2017). *Lactobacillus casei* encapsulated in soy protein isolate and alginate microparticles prepared by spray drying. *Food Technology and Biotechnology 55(2)*, 173–186.

Kjesbu, J. S., Zaytseva-Zotova, D., Sämfors, S., Gatenholm, P., Troedsson, C., Thompson, E. M., Strand, B. L. (2022). Alginate and tunicate nanocellulose composite microbeads – Preparation, characterization and cell encapsulation. *Carbohydrate Polymers, 286*, 119284.

Krunic, T. Z., Bulatovic, M. L., Obradovic, N. S., Vukasinovic-Sekulic, M. S., and Rakin, M. B. (2016). Effect of immobilisation materials on viability and fermentation activity of dairy starter culture in whey-based substrate. *Journal of the Science of Food and Agriculture, 96*, 1723–1729.

Krunić, T., Obradović, N., Bulatović, M., Vukašinović-Sekulić, M., Trifković, K. and Rakin, M. (2017). Impact of carrier material on fermentative activity of encapsulated yoghurt culture in whey based substrate. *Hemijska Industrija 71(1)*, 41–48.

Krunic T. (2017). *Production and application of bioactive proteins and peptides from whey*, A thesis of Ph.D. Department of Biochemistry engineering and biotechnology, Faculty of Technology and Metallurgy, University of Belgrade.

Krunic, T., Rakin, M., Bulatovic, M. and Zaric, D. (2018). The contribution of bioactive peptides of whey to quality of food products. In *Food processing for increased quality and consumption* edited by A. M. Grumezescu, and A. M. Holban, 251–285 Elsevier Inc.

Krunic, T., Obradović, N. and Rakin, M. (2019). Application of whey protein and whey protein hydrolysate as protein based carrier for probiotic starter culture, *Food Chemistry 293*, 74–82.

Krunic, T. (2021). Probiotics in functional food: Impact on immunity and future perspectives. In *Probiotics and their role of health and disease* edited by Y. Olsen, 73-122 Nova Science Publisher.

Krunic, T. and Rakin, M. (2021). FTIR analysis of protein/peptide-based biopolymer used forprobiotic encapsulation. In *Probiotics and their role of health and disease* edited by Y. Olsen, 211-234 Nova Science Publisher.

Krunic, T. and Rakin, M. (2022). Enriching alginate matrix used for probiotic encapsulation with whey protein concentrate or its trypsin-derived hydrolysate: impact on antioxidant capacity and stability of fermented whey-based beverages, *Food Chemistry 370*, 130931.

Liu, H., Gong, J., Chabot, D., Millerc, S. S., Cui, S. W., Zhong, F., and Wang, Q. (2018). Improved survival of Lactobacillus zeae LB1 in a spray dried alginate-protein matrix. *Food Hydrocolloids 78*,100-108.

Lopes, L. A. A., Pimentel, T. C., Carvalho, R. S. F., Madruga, M. S., Galvão, M. S., Bezerra, T. K. A., Barão, C. E., Magnani, M., and Stamford, T. C. M. (2021). Spreadable goat Ricotta cheese added with *Lactobacillus acidophilus* La-05: Can microencapsulation improve the probiotic survival and the quality parameters?. *Food Chemistry 346*, 128769.

Market Research Report, Food supplements — probiotics market (2020–2027) Available online: https://www.fortunebusinessinsights.com/industry-reports/probiotics-market-100083 (2020) accessed on 15 March 2022.

Mirmazloum I, Ladányi M, Omran M, Papp V, Ronkainen VP, Pónya Z, Papp I, Némedi E, and Kiss A. (2021). Co-encapsulation of probiotic *Lactobacillus acidophilus* and *Reishi* medicinal mushroom (Ganoderma lingzhi) extract in moist calcium alginate beads. *International Journal of Biological Macromolecules 192*, 461-470.

Obradovic, (2019). *Characterization and application of natural hydrogels for probiotic starter culture*, A thesis of Ph.D. Faculty of Technology and Metallurgy, University of Belgrade.

Pérez-Luna VH, and González-Reynoso O. (2018). Encapsulation of Biological Agents in Hydrogels for Therapeutic Applications. *Gels 4(3)*, 61.

Petzold, G., Moreno, J., Gianelli, M. P., Cerda, F., Mella, K., Zúñiga, P., and Orellana-Palma, P. (2018). "Food Technology Approaches for Improvement of Organoleptic Properties Through Preservation and Enrichment of Bioactive Compounds," In *Food processing for increased quality and consumption* edited by A. M. Grumezescu, and A. M. Holban, 67–92 Elsevier Inc.

Rajam, R., Karthik, P., Parthasarathi, S., Joseph, G. S., Anandharamakrishnan, C. (2012). *Lactobacillus plantarum*, bile, cell viability, freeze drying, freezing, gastrointestinal system, microencapsulation, probiotics, sodium alginate, spray drying, survival rate, water content, whey protein isolate. *Journal of functional foods,* 891-898.

Ramos, O. L., Pereira, R. N., Rodrigues, R., Teixeira, J. A., Vicente, A. A., & Malcata, F. X. (2014). Physical effects upon whey protein aggregation for nano-coating production. *Food Research International 66,* 344–355.

Yonekura, L., Sun, H., Soukoulis, C., and Fisk, I. (2014). Microencapsulation of *Lactobacillus acidophilus* NCIMB 701748 in matrices containing soluble fibre by spray drying: Technological characterization, storage stability and survival after in vitro digestion. *Journal of Functional Foods 100,* 205-214.

Vivek, K., Mishra, S., and Pradhan, R. C. (2020). Characterization of spray dried probiotic Sohiong fruit powder with *Lactobacillus plantarum. Lebensmittel-Wissenschaft und - Technologie- Food Science and Technology 117,* 108699.

Wang, A., Lin, J., and Zhong, Q. (2020). Physical and microbiological properties of powdered *Lactobacillus salivarius* NRRL B-30514 as affected by relative amounts of dairy proteins and lactose. *Lebensmittel-Wissenschaft und -Technologie- Food Science and Technology 121,* 109044.

Zuidam and Shimoni, 2010). Overview of Microencapsulates for Use in Food Products or Processes and Methods to Make Them. In *Encapsulation Technologies for Active Food Ingredients and Food Processing* edited by N. Zuidam and V. Nedovic, 3–29 Springer.

Chapter 6

Structure-Property Relationship of Alginate Polymers and Its Biomedical Applications

**Bishnu Dev Patra, Sweta Behera,
Smrutirekha Mishra, Sankha Chakrabortty
and Shirsendu Banerjee**[*]
School of Chemical Technology, KIIT University Bhubaneswar, Odisha, India

Abstract

Alginate is a well-known biomaterial-based polymer composed mainly of polysaccharides. It comes under the category of a biopolymer having α-L-Guluronic acid and β-D-Mannuronic groups in its backbone. These possess ionic properties due to which they are known as anionic polymers enabling better ionic cross-linking structures. It also has some interesting properties such as pH responsiveness, film-forming nature, biocompatible, biodegradable, and non-toxic in nature which makes it suitable for different biomedical applications. The common biomedical applications are dental implantations, scaffolding, and drug delivery hydrogels. Mostly, due to its gelling and biocompatibility nature, it renders a great interest in biomedical applications recently. Considering its biomedical applications, it highly necessitates understanding its structure-property relationships. Hence, this chapter will give a brief knowledge of the structure-property relationships, film formation, biocompatibility, and toxicity study of alginate and its different biomedical applications.

Keywords: alginates, polymers, cross-linking, biodegradability, toxicity

---

[*] Corresponding Author's Email: shirsendu.banerjee@kiitbiotech.ac.in

In: Properties and Applications of Alginate
Editor: Michael Y. Wilkerson
ISBN: 979-8-88697-371-6
© 2022 Nova Science Publishers, Inc.

1. Introduction

A bio-polymer that occurs naturally is called alginate. It is a particular class of anionic polysaccharide that is frequently taken from brown seaweeds (brown marine algae). Stanford has patented the method for extracting alginate from brown seaweeds. The primary skeletal component of brown sea algae is alginate, which functions similarly to cellulose in terrestrial plants. Alginate gel maintains the fundamental structure, strength, and flexibility that algae need to endure the force of the water where these seaweeds grow. It is located in the cell wall and between intercellular matrices. Depending on where the algae are gathered, the amount of alginate in them might range from 20% to 60% of their dry weight. However, it is known that brown species of algae have an alginate content of roughly 40% alginate of their dry weight. Alginate is a matrix of cations that are bound to alginic acid and is found in brown algae as gels that contain the ions Na, Ca, Sr, Mg, and Be. These ions give the alginate a very stiff structure, and the divalent cations make the alginic acid molecule exceedingly stable.

A binary copolymer comprising mannuronic acid and guluronic acid is alginate. Although some bacteria are capable of producing alginate, for commercial purposes, all alginates are currently isolated from algal sources. Different sectors employ alginate primarily for the gelation, viscosity, and stabiliser capabilities they give the solution or products they are present in.

2. Synthesis of Alginates

Alginates are mainly extracted from the seaweeds of genus *Alaria, Ascophyllum, Lessonia, Macrocystis, Eisenia, Laminaria, Ecklonia, Nereocystis, Sargassum, Cystoseira*, and *Fucus*, with *Ascophyl lumnodosum* and *Macrocystis pyrifera* being the main assets exploited. The sources are shown in Figure 1.

In the midst of the northern and southern hemispheres, a variety of genera are produced, in addition to some of them being grown offshore as illustrated in Figure 1(e) for the manufacturing of alginate. Typically, a species' origins may be traced back to natural resources. The cost of producing these species is on the higher end, despite the fact that certain nations in the northern hemisphere do so. As a result, Laminaria is mostly used in food, with alginate manufacture taking place in cases of excess. Due to their higher alginic acid concentration than those found in warmer seas, brown seaweeds that grow in water as cold as 20°C are typically employed. Additionally, turbulence promotes greater growth than calm seas do [1].

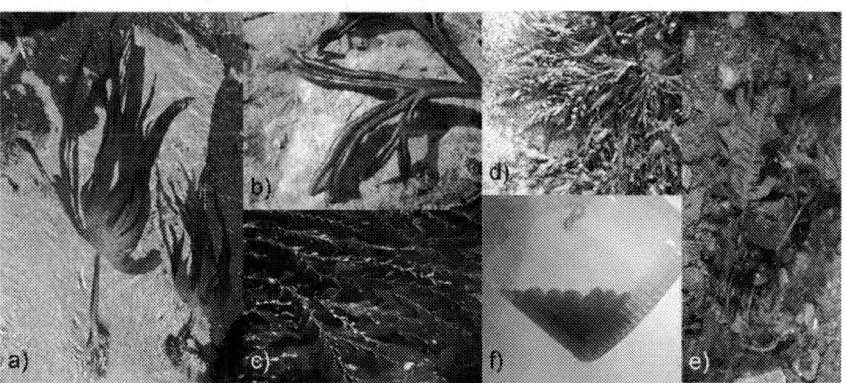

Figure 1. (a) Laminaria ochroleuca from Ínsua beach, afife, north of Portugal; (b) Fucus ceranoides from mondego river estuary (coordinates: 40° 7'31.39"N, 8°46'15.76"W), center-north of Portugal; (c) Macrocystis pyrifera; (d)Ascophyllum nodosum from Praia do norte, Viana Do Castelo, north of Portugal; (e) Alaria esculenta from Eskifjördur, Iceland; (f) Chlorella vulgaris immobilized in calcium alginate gel beads. Adapted from [10.5772/intechopen.88381] (open access).

Seaweeds contain a substance called alginic acid in the form of insoluble mixed salts. By turning the acid into soluble Na and K salts, the extraction is carried out [1]. To separate the components of algae, ion exchange is employed to make them both soluble and insoluble in solvent. The first step is to break down the seaweeds into tiny particles because big compounds are disseminated from plant tissues. As a result, washing, drying, and milling of seaweed constitute the first step. Alginates are dissolved as sodium alginate by reacting with a strong alkali, first pretreating with HCl, then drawing off with Na_2CO_3. Alginate segregation is primarily an ion exchange process. [2, 3]. Alginates can be separated from other soluble components in the crude alginate extract solution using a variety of procedures. For instance, sodium alginate can be precipitated by adding C_2H_5OH [4]. Calcium alginate can also precipitate out when $CaCl_2$ solution is added to a stirrer at a faster rate than HCL, whereas alginic acid can precipitate out when HCL is added. The next step in the removal of alginate is the placid acid treatment, which removes undesirable substances (often HCL) and converts the alginate found in cell walls into alginic acid to achieve the highest extraction efficiency [5]. Alginic acid is primarily found in intercellular mucilage. By neutralizing the alginic acid with either Na_2CO_3 or NaOH, the acid is removed as soluble Na. While the soluble alginate precipitates by converting to alginic acid or ions of Ca or Na, the insoluble residue is residued using filtration, flotation, or centrifugation techniques. By neutralizing the created alginic acid with the appropriate chlorites or hydroxides, it is then converted into the required counter ion. The basis and arrangement of alginate's component parts determine how differently the alginate revival process differs [2].

3. Applications of Alginates

Alginate is a key component of many different businesses, including those that deal with food, cosmetics, paper, agriculture, textiles, etc. Alginate is becoming into a crucial chemical in biomedical applications because of its biocompatibility and low toxicity. It is employed in the pharmaceutical industry for a variety of purposes, including gel formation, stabilizing agents, and drug delivery systems. In terms of quantity, alginate is the most common seaweed polymer utilized by many sectors.

Alginate polysaccharide is utilized as a natural addition in the food and beverage industries primarily because of its stabilizing, emulsifying, and gel-forming properties. They are actively used to stabilize the structure of food in jams, jellies, ice creams, dairy goods, etc. Alginate's key benefit as a food product is that because humans lack the enzymes necessary to break down the alginate molecule, it behaves as a dietary fibre, increasing satiety, lowering calorie intake for people by reducing food intake, and preventing obesity. Alginate is increasingly in demand in the food business on a daily basis. Alginate is a soil conditioner used in agricultural fields. Alginate is utilized as a natural hydrogel for soil agricultural regions since it is a super-absorbent (water retaining material) and can absorb water up to 100 times its own mass. Alginates are a thickening agent for the paste holding the dye used in the textile industry. Alginate can be removed from the dye without causing a reaction and yet produce the desired printed result. Alginate is used for surface sizing in the paper industry. Alginate is used as a film forming in the printing industry and improves the ability and resistance of ink printing. Alginates are frequently utilised in the craft sector. Alginates are utilized in the cosmetics industry as thickening agents, emulsifiers, and consistency enhancers in cosmetic formulas by producing a moisture-retaining surface film. They have good moisture-retentive qualities, which keeps the skin moisturized. Alginates have been discovered to be anti-oxidative agents and can be administered topically to the skin to stop cutaneous disorders and skin ageing. Because alginates are harmless, biodegradable, and derived from natural sources, they are an excellent substitute for other packaging materials. The welding industry uses coatings made of alginate as a flux on welding rods or electrodes.

Alginates are used because of three characteristics: their ability to thicken the resulting solution when dissolved in water, which essentially makes the solution more viscous; their high propensity for gel formation, whereby sodium alginate and water form a gel when calcium salt is added (the formation of gel is due to a chemical reaction where the sodium is displaced by calcium from the alginate while holding the longer molecules together resulting in gel formation); thirdly, alginates form calcium and sodium alginate films along with calcium alginate fibers.

Long chain molecules contained in alginates with various acidity components, particularly M and G, are responsible for these features (for simplicity). Depending on the type of seaweed, the M/G ratio changes. This implies that the traits and qualities of alginates are frequently variable. The amount of diversity in gelling qualities that can be obtained is often determined by this component. In other words, depending on the type of alginate, the viscosity of the solution where alginates are dissolved may change. Similar to that, depending on the type of alginate employed, the gel strength following the addition of calcium salt may also change.

3.1. Textile Printing

Alginates are used in paste thickeners, particularly ones that incorporate colour. The pastes are applied to the fabrics using either screen or roller printing technology. After reactive dyes were introduced, alginates became important thickeners. In the fabrics, the colors chemically combine with the cellulose. The usual thickeners, like starch, react with the dyes and reduce the amount of color produced. In addition, the by-products are difficult to wash off. The higher cost of alginates compared to starch is another drawback. Because of this, manufacturers of starch are working to create starch that won't react with reactive colors, so preserving the market. This use of alginates creates a sizable market despite being impacted by the recession, which frequently causes a fall in demand for fabrics and clothes. Depending on the type of printing machinery, the viscosity ranges for alginates are from medium to high.

3.2. Food

The thickening properties of alginate are used by manufacturers of sauce, syrup, and ice cream toppings. Pie fillings typically thicken when alginates are added, and nonsticky icings make it easier to wrap baked items in plastic. Alginates are also used to thicken mayonnaise and salad dressing. This limits the ability to separate the oil and water from their natural oil-water emulsion properties. Alginate restores the yogurt's firmness, body, and shine, but PGA is also used to stabilize milk proteins under the highly acidic conditions that are typically present in many yoghurts. When the environment is acidic, PGA may be used instead of sodium alginate to keep fruit pulps suspended in various fruit drinks. A phosphate/alginate mixture can also be used to maintain the suspension of the cocoa in chocolate milk. Alginates can also stabilize and thicken whipped cream in very small amounts.

Alginates have a few uses that don't call for viscous or gelling properties. They act as stabilisers for ice cream because the use of alginate reduces the formation of ice crystals after freezing, resulting in an even product. This is especially important since refrozen ice creams lacking alginate develop huge ice crystals, which are often undesirable, when they are transported from the supermarket to the home freezer. Alginates also aid in preventing the ice cream from melting more quickly. Propylene glycol alginate, which is used in very small amounts, gives beer a long-lasting foam that is believed to be a sign of high beer quality. Similar to wine clarification, sodium alginate is also used in wine fining. Since the creation of artificial cherries, the gelling properties of alginate have been utilized.

3.3. Pharmaceutical and Medical Uses

3.3.1. Pharmaceutical Applications

Because alginates play a significant role in controlled release medications, their typical functions in the pharmaceutical industry include thickening, gel formation, and stabilising. The oral dose is now the most common way that alginate is used in pharmaceuticals, although alginate hydro-gel is increasingly being used as a repository for tissue-specific drug delivery. Some of the recent progress of the same is shown below.

3.3.2. Delivery of Small Chemical Drugs

In particular, when secondary or primary bonds between alginates and medicines can be employed to control the entire kinetic process of the release of drugs, alginate gels are likely highly relevant for a number of low MW drug delivery methods. Alginate gel is typically nanoporous, with pores that are approximately 5 nm in size [6], which causes small molecules to disperse quickly throughout the gels. For instance, flurbiprofen is largely released from cross-linked, partially oxidised alginates gel after around 90 minutes. However, amalgamation of Ca ion and adipic Acid Dihydrazide (combination of ionic and covalent cross-linking) into a bead made from incompletely oxidised alginates results in a lengthened release because of the increased number of cross-links and consequently reduced enlargement. [7]. Alginates gel that has partially oxidised can also be used to administer antineoplastics locally and under control. Since compound medications' chemical makeup and method of integration can significantly alter the kinetic characteristics, compound pharmaceuticals may be put inside alginate-based gel for concurrent or chronological administration.

3.3.3. Protein Delivery

Due to the development of recombinant DNA technology, the market for protein treatments is rapidly expanding and a variety of protein medications are now available. Alginates are excellent choices for protein medication administration because protein can be added to an alginate-based formulation under relatively benign circumstances, avoiding denaturation, and the gel will prevent the protein from deteriorating before release. The protein release rate from alginate gels is managed using a variety of techniques. Because of its inherent porosity and hydrophilic nature, alginates gel typically releases protein at a rapid rate. However, heparin-binding enlargement factors like as vascular endothelial growth factor (VEGF) or basic fibroblast growth factor (bFGF) exhibit comparable, reversible binding to alginate hydrogels allowing a continuous and contained release [8, 9]. By altering the rate at which the gel is depleted in this case (e.g., by using partially oxidised alginate), the discharge can be intentionally affected, making the release of the protein in any case partially dependent on the degradation reaction [10]. Progress has been made in controlling the release of angiogenic compounds from alginates gels, especially for variables other than heparin binding.

High-pH proteins like chymotrypsin and lysozyme are successfully encapsulated in ionically cross-linked alginates micro-spheres; the proteins tend to cross-link the sodium alginate, allowing for more sustained release. [11].

Without the need for a catalyst, poly ((2-dimethylamino) ethyl methacrylate can also react with alginate that has been oxidized. Additionally, gel beads can be created by dripping the alginate derivative solution into an aqueous calcium chloride solution to create protein delivery particles [12]. Alginates can also be utilized as building blocks in the synthesis of a tetrafunctional, acetal-linked network of polymers designed to create gels with pore sizes that can be adjusted and react to stimuli. The gel may release the loaded protein in an almost neutral pH with zero order kinetics while also protecting acid labile proteins like insulin from denaturation in stomach circumstances (pH = 1.2) [13].

Numerous cross-linking or encapsulating techniques, as well as improving contacts between proteins and hydrogels, may be used to address the limited encapsulation efficacy and rapid discharge from alginates gel demonstrated by many proteins [14]. For instance, alginates micro-spheres containing insulin have been created by mixing alginates with anionic polymers (such as polyphosphate, cellulose acetate phthalate, and dextran sulphate), followed by the addition of chitosan coating to prevent the release of insulin at gastric pH levels until it reaches intestinal pH [15]. By depositing layers of Bombyx mori silk fibroin over alginates microspheres, a durable shell can be created that also acts as a barrier to the amalgamated protein's ability to diffuse [16].

Protein discharge is further facilitated by a mixture of microspheres acting as a depository for alginate hydrogels and protein. Poly(d,l-lactide-co-glycolide) (PLGA) microsphere suspension in alginate combined with ionic cross-linking can be used to create hydrogels that are microsphere-loaded. The release of bovine serum albumin (BSA), a dummy protein, from this delivery arrangement can primarily be changed by adjusting the ratio of mixing between alginate hydrogel and PLGA microspheres, free of total BSA content, and the volume of PLGA microspheres used [17]. Uniform dispersion of PLGA microspheres can be achieved by SEM. By altering the ratio of microspheres to gels, TAT-HSP27 (heat shock protein 27 fused with transcriptional activator) discharge performance may also be synchronised [18].

3.3.4. In Pharmaceutical Industries

Pharmaceutical businesses use alginates for a variety of purposes due to their properties, including gel formation, biocompatibility, and non-toxicity.

Alginates are traditionally employed in pharmaceutics as a thickening, stabilizing, and gel-forming agent. Alginates can also be utilized for medication products with controlled release. Currently, oral dosage versions of alginate are used in medicinal applications most commonly. A number of low molecular weight pharmaceuticals are transported using alginate hydrogel in drug delivery systems, and it is more efficient and practical to use the primary or secondary connection between the drug molecule and the alginate molecule to control the drug's kinetics. The majority of small medication molecules quickly diffuse through alginate gels because they have pores with a diameter of less than 5 nm. For instance, the anti-inflammatory drug Flurbiprofen virtually releases from alginate gel in 1.5 hours. For controlled and targeted medication administration, particularly the delivery of neoplastic agents to malignant cells, alginate gels that have partially oxidized are used.

Alginate-based gels can contain various medications for simultaneous and sequential drug administration. Contrary to doxorubicin, which was covalently bonded to alginate and was released via chemical hydrolysis of the cross-linked alginate gel, methotrexate, an immunosuppressant, does not interact with alginate and is released rapidly by diffusion.

Using amphiphilic alginate gel particles, hydrophobic drugs' release can be regulated. Theophylline, a drug with weak water solubility, can be distributed in a regulated manner by grafting alginate with poly-caprolactone (PCL) and then cross-linking with calcium ion. Theophylline can be released over time by combining alginate with carbon nanotubes. Since alginate gels have greater mechanical stability and no detectable cytotoxicity while maintaining the structure and shape of the alginate microspheres, they can be used as an acceptable delivery vehicle to the colon and intestinal tissue. Alginate and chitosan, a sugar produced from animal exoskeletons, interact to form ionic compounds. Complicated ionotropic gelation procedures are employed to produce multi-particulate systems of alginate and chitosan-containing triamcinolone for colonic medication administration. Albendazole can be physically captured and transported into the digestive tract using magnetic alginate chitosan beads. ATRA (all-trans retinoic acid) can be released into the skin over an extended period of time by using alginate microparticles that have been coated with chitosan. Metronidazole is also ionotropically gelled with alginate beads that have been chitosan-treated. Additionally, alginate gels are used to make a matrix into which small drug delivery systems can be placed. Alginate gels and chitosan nanoparticles that have been loaded with amoxicillin can be used to treat H. pylori infection. Because the outer layer of the alginate gel protects the amoxicillin-loaded nanoparticles from the gastrointestinal environment, amoxicillin can only interact with cells intracellularly, where H. pylori infections occur. Recombinant DNA technology is developing thanks to the availability of many protein medications. Alginate is regarded as the best molecule for protein drug delivery because, when included with alginate-based formulations under relatively mild circumstances, protein medicines do not denature or break. The protein medicine molecule is additionally shielded from degradation by the alginate gel until it is released.

Alginate gels often release protein quickly because of their hydrophilic nature. However, sustained and targeted protein drug release from alginate gels can be achieved via reversible binding of heparin-binding growth factors like VEGF (vascular endothelial growth factor) or bFGF with alginate gels. The rate of medicine release can be easily regulated by modifying the rate at which alginate gels break down. Ironically, high-pI proteins like chymotrypsin and lysozyme can be physically cross-linked with sodium alginate for prolonged release. This allows high-pI proteins to be encapsulated in cross-linked alginates. Alginate can be utilised to make stimuli-responsive gels that release proteins when they are required while avoiding the stomach environment's tendency to denaturize proteins like insulin. The low encapsulation effectiveness and rapid release of protein from alginate gels can be addressed using a variety of cross-linking and encapsulation techniques as well as improved protein-hydrogel interactions. Alginate, for instance, can be combined with an anionic polymer to produce microspheres that are filled with insulin and covered in chitosan to shield it from the stomach environment. Alginate microspheres can also be coated with Bombyxmori silk fibroin to give the protein-encapsulated shell a more durable surface.

3.3.4.1. Wound Dressing

Alginate-based wound dressing also delivers moisture to the site to speed up healing, whereas traditional wound dressing merely keeps the wound dry and inhibits the entry of any germs. In order to maintain a physiologically moist environment at the wound site, prevent bacterial infection, and eventually hasten the healing process, alginate gel dressing has the capacity to absorb wound fluid, dry it out for re-geling, and then supply moisture to the dry wound. These procedures can also encourage the growth of granulation tissue and hasty healing. Alginate gel, which is then processed to produce fibrous non-woven dressings and freeze-dried porous sheets, is the conventional method for making alginate wound dressings. The gel is produced by ionic cross-linking an alginate solution with calcium ions. Only a few examples of commercially marketed alginate dressings include Algicell, Algisite, Comfeel Plus, and others. Additionally, research has been done on a number of more practical bio-active wound dressings. It has been found that partially oxidized alginate gels can speed up wound healing by releasing di-butyryl cyclic adenosine monophosphate, which regulates the proliferation of human keratinocytes, in a controlled manner. Complete healing takes about 10 days. Silver can be added to alginate-based dressings to improve their antibacterial activity and binding affinity to other chemicals. Adding silver to alginate dressings can also boost their antioxidant power. Because zinc ions have antibacterial and immune-stimulating qualities as well as the capacity to increase the growth factor of adjacent cells, alginate fibers that have been cross-linked with zinc ions are being studied for use in wound dressings.

3.3.4.2. Cell Culture

Alginate gels are used as a model system for mammalian cell development in biomedical research. These gels can be utilized with 2D and 3D culture methods. Alginate lacks mammalian cell receptors and has low protein adsorptions to alginate gels, making it an excellent culture media for highly selective and quantitative modes of cell adhesion procedures, such as coupling of synthetic peptides specific for cellular adhesion receptors. RGD-modified alginate gels are frequently utilized as substrates for in vitro cell growth. RGD peptides are found in alginate gels, which assist control the characteristics of interacting cells. For instance, the chemical conjugation of RGD peptides to alginate gels significantly enhances the adhesion and proliferation of myoblasts cultivated on alginate-based culture media when compared to alginate gels that have not been altered.

3.3.4.3. Tissue Engineering

For the transplanting of cells during tissue engineering, alginate gels are thought to be excellent materials. Alginate hydrogels are used to transfer proteins or cells (regenerative agents) that can hasten or facilitate the body's production of new tissues and organs. Because the pore size of alginate gels is about 5 nm, there are certain size constraints for the regeneration agents that can be released by diffusion from alginate hydro-gels. Molecules that are too large to diffuse out of the alginate gel, however, might also be released when or if the gel dissolves. Alginate gels are regarded as effective angiogenic molecule delivery systems. It is possible to induce neovascularization (the growth of new blood vessels) by introducing different cell types into the body, administering angiogenic drugs, or doing both. Like VEGV, heparin-binding growth factor is gradually released locally. Because bone cells don't heal well, there are some limitations on how bone injuries can be treated. Alginate gels can be utilized to restore bone tissues by delivering osteoinductive factors and bone-forming cells to the injured bone tissue. Alginate gels are also utilised to engineer and regenerate a variety of issues with the liver, skeletal muscles, pancreas, lungs, and other organs. Alginate gels are preferred over other materials for bone and cartilage tissue regeneration due to additional advantages such as the ability to fill irregularly shaped defects and ease of chemical modification with adhesion ligands (like RGD) and controlled release tissue induction factors. Gels made of alginate can also be injected into the body with little discomfort.

Orthopedics still has a very difficult time repairing injured cartilage. Chondrogenic cell transplantation has shown effective for the restoration and repair of damaged cartilage when done using alginate gel. The regeneration and engineering of numerous different tissues and organs, such as the liver, pancreas, nerves, skeletal muscle, and nerve endings, can be effectively regulated by alginate gels. These techniques can be carried out by alginate gels, it has been found. Current methods for regenerating skeletal muscle include growth factor administration, cell transplantation, or both. VEGF and IGF-1 delivery via alginate gels, which are utilized to regulate both angiogenesis and myogenesis. The targeted and continuous delivery of angiogens and myogens results in improved muscle regeneration and functional muscular forms. Using alginate gels, the central and peripheral neural systems have been successfully restored. Alginate hydrocolloids are used in dental medicine as an irreversible footprint material to produce high-quality imprints that precisely mimic a footprint. The main advantages of alginate hydrocolloids are their low cost, lack of side effects, and ease of modification.

Conclusion

Alginate is a type of biomaterial that is now being studied and used for several biomedical applications. Because of its bio-compatibility, alginate and its derivatives are currently among the most widely used bio-polymers.

Alginate hydrogels are widely employed in biomedical applications, including tissue engineering, cell culturing, dental dentistry, and controlled drug delivery. Alginate hydrogels are safe to administer to the chosen human body since they are biocompatible and structurally equivalent to the macromolecules of the human body. It is crucial to utilise and evaluate alginate that is as pure as feasible for biomedical applications since impurities can damage the material's biocompatibility. As a pharmaceutical ingredient and material for wound dressings, alginate is regarded as safe for use in clinical settings. Alginate and its derivatives are used in numerous biomedical applications without damage.

References

[1] Clare K. Algin. *Industrial Gums.* 1993:105-143.
[2] McHugh D J. Production and Utilization of Products from Commercial Seaweeds. *FAO FisheriesTechnical Paper No.* 288. Vol. 1891987.
[3] Sime W J. Alginates. In: Harris P, editor. *Food Gels.* Dordrecht: Springer Netherlands; 1990. pp. 53-78.
[4] Haug A, Smidsrød O. Strontium–calcium selectivity of alginates. *Nature.* 1967; 215(5102):757.
[5] Pereira L, Gheda S F, Ribeiro-Claro P J. Analysis by vibrational spectroscopy of seaweed polysaccharides with potential use in food, pharmaceutical, and cosmetic industries. *International Journal of Carbohydrate Chemistry.* 2013;2013(vi):1-7.
[6] Boontheekul T, Kong H J, Mooney D J. Controlling alginate gel degradation utilizing partial oxidation and bimodal molecular weight distribution. *Biomaterials* 2005;26:2455–65.
[7] Maiti S, Singha K, Ray S, Dey P, Sa B. Adipic acid dihydrazide treated partially oxidized alginate beads for sustained oral delivery of flurbiprofen. *Pharm. Develop. Technol.* 2009;14:461–70.
[8] Lee K Y, Peters M C, Mooney D J. Comparison of vascular endothelial growth factor and basic fibroblast growth factor on angiogenesis in SCID mice. *J. Control Release* 2003;87:49–56.
[9] Silva E A, Mooney D J. Effects of VEGF temporal and spatial presentation on angiogenesis. *Biomaterials* 2010;31:1235–41.
[10] Wells L A, Sheardown H. Extended release of high pI proteins from alginate microspheres via a novel encapsulation technique. *Eur. J. Pharm. Biopharm.* 2007; 65:329–35.
[11] Gao C M, Liu M Z, Chen S L, Jin S P, Chen J. Preparation of oxidized sodium alginate-graftpoly((2-dimethylamino) ethyl methacrylate) gel beads and in vitro controlled release behavior of BSA. *Int. J. Pharm.* 2009;371:16–24.
[12] Chan A W, Neufeld R J. Tuneable semi-synthetic network alginate for absorptive encapsulation and controlled release of protein therapeutics. *Biomaterials* 2010;31:9040–7.
[13] Silva C M, Ribeiro A J, Ferreira D, Veiga F. Insulin encapsulation in reinforced alginate microspheres prepared by internal gelation. *Eur. J. Pharm. Sci.* 2006;29: 148–59.

[14] Wang X, Wenk E, Hu X, Castro G R, Meinel L, Wang X, Li C, Merkle H, Kaplan DL. Silk coatings on PLGA and alginate microspheres for protein delivery. *Biomaterials* 2007;28:4161–9.

[15] Lee J, Lee K Y. Injectable microsphere/hydrogel combination systems for localized protein delivery. *Macromol. Biosci.* 2009;9: 671–6.

[16] Lee J, Tan C Y, Lee S K, Kim Y H, Lee K Y. Controlled delivery of heat shock protein using an injectable microsphere/hydrogel combination system for the treatment of myocardial infarction. *J. Control Release* 2009;137:196–202.

[17] Soon-Shiong P, Heintz R E. Insulin independence in a type 1 diabetic patient after encapsulated islet transplantation. *Lancet* 1994;343:950.

[18] Lee J, Lee K Y. Injectable microsphere/hydrogel combination systems for localized protein delivery. *Macromol. Biosci.* 2009;9: 671–6.

Chapter 7

Properties and Medical Applications of Alginate

Seyed Rasoul Tahami, PhD
and Nahid Hassanzadeh Nemati*, PhD
Department of Biomedical Engineering, Science and Research Branch,
Islamic Azad University, Tehran, Iran

Abstract

Alginate, which refers to all derivatives of Alginic acid, is a natural anionic exopolysaccharide commonly obtained from brown seaweed or certain species of bacteria called Pseudomonas and Azotobacter. The structure of alginate is heteropolymeric, i.e., it consists of two types of Uronic acid, including β-D-Mannuronic acid (M) and α-L-Guluronic acid (G), which are linked by 1,4-glycosidic bonds. The number and length of M blocks and G blocks and MG blocks in the polymer structure vary significantly between alginates and affect their physical and chemical properties. The most important feature of alginates that makes it possible to use them in various applications, especially biomedical applications, is the formation of hydrogels, which in turn is dependent on divalent cations and their bonding and the formation of crosslinking structures. Due to their unique physicochemical properties, they are widely used in tissue engineering, drug delivery systems, advanced wound dressings, bioadhesives, cosmetics, food, and agriculture, the range of their use is increasing and expanding day by day. This chapter, after reviewing how alginate is discovered, describes the main properties of alginate, discusses the sources and chemical structure of alginate, and its various applications, especially in drug release and tissue engineering.

---

* Corresponding Author's Email: nahid_hasanzadeh@yahoo.com.

In: Properties and Applications of Alginate
Editor: Michael Y. Wilkerson
ISBN: 979-8-88697-371-6
© 2022 Nova Science Publishers, Inc.

Keywords: alginate, alginic acid, seaweed, azotobacter, hydrogels, tissue engineering, and drug delivery systems

Introduction

From the beginning, human has used various materials to replace limbs or parts of limbs: glass for the eyes, wood for the teeth, and so on. The Romans, Chinese, and Aztecs used gold in dentistry. The Egyptians and Indians used linen for suturing (as well as horse hair, etc.). These were the materials they had access to and used in their daily lives (Migonney, V. 2014).

Biomaterials that significantly affect human health, the economy, and the scientific community are about 70 to 80 years old. Biomaterials and medical equipment made from them are commonly used in cardiovascular, orthopedic, dental, ophthalmic and reconstructive, surgical surgeries, bioadhesives and controlled drug delivery systems. The compelling and humane aspect of biomaterials is that millions of lives have been saved and the quality of life of millions more has been improved based on devices made from biomaterials (Buddy D. Ratner et al.'s, 2020).

Biomaterial for each application must be:

- Biofunctional
- Biocompatible
- Resistant to corrosion
- Biocompatible
- Blood compatible
- Non-toxic
- Non-inflammatory
- Non-Nonpyrogenic
- Non-allergenic

Must also:

- Allow sufficient long-term use without rejection by the host organism or loss of desired function.
- Biomaterials should also consider the structure and function of surrounding tissues and organs, so they should not disrupt the

structure and function of surrounding tissues and organs (Andrzej Hudecki et al.'s, 2020).

Migonney, V. 2014 defined "biomaterials" as "Materials intended to repair or replace all or part of a damaged tissue or organ." Therefore, biomaterials are different from other substances in that they should not cause host immune response, toxicity, allergic and inflammatory reactions in contact with tissues and organs of the living system, meaning that they are biocompatible and blood compatible and should not be rejected.

Alginate is a natural anionic and hydrophilic polysaccharide. It is biosynthesized from the most abundant biomaterials and is obtained from brown seaweed and Pseudomonas bacteria and Azotobacter (Sun et al.'s, 2013; Rinaudo, Marguerite 2013). Alginate is a linear, water-soluble copolymer consisting of two types of blocks: units of glucuronic acid (block G) and manuronic acid (block M) irregularly arranged in blocks GG, MG and MM, the ratio and pattern of which depend on their origin. The most desirable properties of alginate are biocompatibility, low toxicity, relatively low cost and its gentle and easy gelation. Alginates in the presence of divalent cations (mainly calcium cations) very quickly form a hydrogel polymer network with a three-dimensional configuration that is able to absorb large amounts of water (Moore, A. 2015). Due to its ability to sol/gel transition and form semi-solid and solid structures, Alginates is commonly used as a viscosity enhancer in the food and pharmaceutical industries, as a structural support biomaterial for tissue regeneration (teeth, bones and cartilage). Alginates have been shown to increase the bioavailability and immunogenicity of antigens, so they have also been studied as vaccine adjuvants (Szekalska et al.'s, 2016). In this review, the properties of alginate and its applications in biomedical and medical sciences will be discussed.

Alginate Discovery

Alginate was first discovered by Stanford, an English chemist, in 1883. She isolated a brown paste with 2% calcium carbonate from brown algae and then acidified solution to precipitate this substance, which she called "algin." In 1928, Nelson and Kercher (Mellon Industrial Research Institute, Pittsburgh, Pennsylvania) obtained purified alginic acid from Laminaria agardhii and Macrocystis pyrifera, which were made from polyuronic acids. Alginate production from bacteria was first reported by Linker and Jones in 1964. In the same year, Gorin and Spencer reported the production of alginic acid by Azobacter vinelandii. The first successful company in the production of pure alginates on a large and commercial scale, Kelco Company, was in San Diego, USA, in 1929. Later, Merck and Co. Inc., USA became the largest alginate producers in the world by purchasing two of the largest alginate producers, i.e., Kelco (USA) and Alginate Industries Ltd (England) in 1972 and 1979 (Moradali, M. et al.'s, 2018).

General Properties of Alginate

Alginates (ALGs) are anionic polysaccharides that are naturally present in the cell wall structure of brown algae such as *Ascophyllum nodosum*, *Laminaria hyperborea*, and *Macrocystis pyrifera*. They are also present as a capsular polysaccharide in bacterial strains such as *Azotobacter* and *Pseudomonas*. They are present in the cell wall of brown algae as calcium, magnesium and sodium salts, therefore, it is usually called "alginic acid and its salts" (Shakeel, A. 2019).

Extraction of alginate from algae sources involves various stages in which brown algae are mechanically harvested and dried in the first stage. After drying and milling algal material, it is treated with mineral acids to remove counterions by proton exchange. In the second step, insoluble alginic acid is dissolved by the neutralization process or by the use of sodium hydroxide (NaOH) or sodium carbonate (Na2CO3) as the alkali, and sodium alginate is obtained. At this stage, alginate contains several cytotoxic impurities that make them unsuitable for biomedical applications. Further purification is required to remove these impurities. Another method for extraction is using Ba2 + ions, because it has a higher affinity towards alginate than Ca2 + ions and the resulting product, Ba-alginate gels, are stable at neutral and acidic pH but decompose at alkaline pH (Hasnain, Md. et al.'s, 2019).

Structure and Characterization

Alginate is an unbranching linear biopolymer. The alginates consist of 1,4-β-D-mannuronic acid (M) and 1,4 α-L-guluronic acid (G) monomers, which are either homogeneous, that is composed of M blocks or G blocks, or are heterogeneous, composed of M and G blocks (MG). Each alginate species, depending on the source, may have differences in the ratio, composition, and sequence of manuronic and guluronic acid blocks that cause differences in their properties. Therefore, the ratio of these two acids varies from species to species and also to different parts of the seaweed. (Pereira, L. and Cotas, J. 2020).

Molecular Weight and Solubility

The molecular weight of commercial sodium alginate is between 32 and 400 kDa (Saji, S. et al.'s, 2022). A mixture of high molecular weight and low molecular weight alginate polymers is used to adjust the viscosity of the alginate solution. Viscosity in alginate reacts to pH changes, so that as the pH decreases, the viscosity increases and reaches a maximum at PH = 3-3.5, because the carboxylate groups are protonated in the alginate backbone and form hydrogen bonds (Reddy, S. 2021).

Biocompatibility

Alginates have been widely used in biomedical and pharmaceuticals due to their non-toxicity, biocompatibility, non-immunogenicity, hydrophilicity and biodegradability. Researchers have reported that the presence of mitogens (a known contaminant class of alginates produced during alginate extraction) in raw alginates is significantly responsible for the immunogenic effects. The US Food and Drug Administration (FDA) considers alginates to be a "Generally Referred as Safe" (GRAS) substance. Block M is known as the main initiator of immune system reaction to foreign body (Zhang, H. et al.'s, 2021; Hasnain, Md. et al.'s, 2020).

Hydrogel Formation

Alginate hydrogels are easily obtained by using a suitable cross-linking ion (divalent cation, i.e., Ca 2+), which induces 3D network formation by chelation by G sequences in the alginate chain. The ionic binding ability of alginate is the main property controlling hydrogel formation. An alginate gel is a water-swollen continuous network in which physical cross-links hold different polysaccharide chains together. The main limitation in using alginate gel is being in an environment that has compounds capable of chelating divalent ions, as well as in an environment with a high concentration of competing ions. There are several strategies to overcome the limited stability of alginate gels under physiological conditions. As a first option, the replacement of calcium ions with stronger binding ions, such as Ba 2+ or Sr 2+, as well as the use of alginate chains with a high proportion of G have been reported. Thus, a significant increase in the stability of alginate hydrogel was observed and although the toxicity of these ions used for biomedical applications is still a concern. Another method for stabilizing alginate gels involves covalent cross-linking in addition to physical (ion-induced) bonds. Various chemical techniques have been performed, including covalent linking of alginate with synthetic polymers, a combination of covalent and ionic crosslinking on polysaccharide, and direct crosslinking of poly(l-lysine) on alginate (Donati, I. and Paoletti, S. 2009).

Medical Applications

Alginate has been widely studied in tissue engineering as a biological material for the regeneration of skin, cartilage, bone, and heart tissue. Also, there is a wide scope for the application of alginate in the biomedical and pharmaceutical industry, including wound healing, Cell attachment, immobilization, and matrix for living cells, delivery of bioactive agents, such as chemical drugs and proteins; and formulation of tablets, to promote greater protection, drug stabilization, and sustained drug release. Alginates also have wide applications in several industrial fields including textiles, food, and cosmetics. It is noteworthy that there are several commercial products based on alginates (Gheorghita, P. et al.'s, 2020; Pahlevanzadeh, F. et al.'s; Dong et al.'s, 2013).

Drug Release System

Alginates have attracted much attention as a potential drug release control device (hydrophilic and hydrophobic). Alginate has been successfully used as a matrix for the entrapment and/or delivery of drugs and bioactive molecules such as proteins or growth factors. Encapsulation and release of proteins from alginate gels can increase their efficiency and release. Molecules are encapsulated in polymers for protection and controlled release. They have many advantages such as the following: they protect the gastric mucosa from the aggressive effect of the drug or protect acid-sensitive drugs from gastric juice, they provide controlled drug delivery and when administered orally They are non-toxic. Drugs with unfavorable solid-state properties (e.g., low solubility) also benefit from encapsulation.

Today, solid preparations based on alginates are available, such as oral tablets, microcapsules, implants, and topical delivery systems. The excellent properties of alginate have led to its use in a wide range of pharmaceutical applications. Alginate has been used as a biological material for drug delivery in the form of microcapsules, microparticles, gel particles, pellets, and seeds. Alginate can be used to modify drug release profiles during oral controlled drug delivery or targeted delivery and controlled release of bioactive compounds (Farokhi, M. et al.'s, 2020).

Tissue Engineering

Cartilage and Bone Tissue Engineering

Alginate biomaterials have been used in the treatment of bone injuries as delivery systems for bone-inducing factors, drugs, osteogenic cells, or a combination of both. Compared to other materials, alginate gels have advantages for bone and cartilage regeneration, because they can enter the body in a minimally invasive process, they have the ability to fill defects (with regular or irregular shapes), and they are easily modifiable. chemically with adhesive ligands (eg, RGD), and can control the release of drug/bioactive molecules into tissues (e.g., BMP, TGF). However, alginate gels do not have sufficient mechanical strength, but this feature can be overcome when combined with other components such as hydroxyapatite, calcium phosphate, or chitosan. For example, combining alginate gels with collagen type I and tricalcium phosphate improve cell adhesion and proliferation that do not readily adhere or proliferate on pure alginate gels. Alginate/chitosan gels that entrapped MSCs and growth factors (BMP-2, BMP-7, and VEGF) also showed trabecular bone formation potential in animal models.

Alginate gels are potential biomaterials that have proven useful for grafting chondrocytes to restore damaged cartilage in animal models. For example, some studies used the suspension of chondrocytes in alginate solution mixed with calcium sulfate and injection into facial implant molds to produce cartilage (Rastogi, P. and Balasubramanian, K. 2019).

Cardiac Applications

Alginate-based biomaterials have also been developed for use in cardiac applications. Alginate was described as a material of non-thrombotic nature, which was confirmed by several biomedical applications and clinical assays. Therefore, it can be an attractive candidate for cardiac applications such as heart valve tissue engineering. For example, research has shown that alginate/collagen composite microparticles can provide a physiological microenvironment for the heart and blood vessels that can be suitable for the growth of cardiac-like tissues. Other biomaterials, such as injectable hydrogels, are for use in the repair. The heart was tested. They have the potential to deliver bioactive molecules, therapeutic agents, and cells in situ in damaged tissue to regenerate functional cardiac tissue. The hydrogel is injected into the damaged heart muscle and polymerizes in place between the cells and fibers of the damaged tissue. This biomaterial provides the necessary physical support for the heart muscle during the repair process. Sometimes, 3D alginate scaffolds can be modified with RGD peptide to improve cell attachment and growth and increase angiogenic growth factor expression. Also, the stabilization of alginate biomaterials with RGD peptide has been presented as an important parameter in the repair of functional heart muscle tissue and in improving the protection of the regenerated tissue in culture (Rastogi, P. and Balasubramanian, K. 2019).

Wound Dressing

Usually, wound dressings are synthesized from synthetic polymers and biopolymers. When used as a wound dressing, sodium alginate is usually combined with calcium chloride to form a pad or rope. Ca2+ from the dressing interacts with Na+ in the wound fluid, so that the fiber of the dressing swells and partially turns into a gel that moisturizes the wound bed and accelerates the healing process (Soorbaghi, et al.'s, 2019).

Tahami et al.'s (2020), in order to obtain Polyvinyl alcohol/Sodium alginate electrospun nanofibrous mats containing herbal medicine with uniform distribution and controllable structural properties with the aim of biomedical applications, the effect of electric potential on the morphology (morphology) of polyvinyl alcohol/sodium alginate and polyvinyl alcohol/sodium alginate/Calendula officinalis electrospun nanofibers examined.

In another work, the ability of alginate/polyvinyl alcohol containing Calendula officinalis extract in wound healing was investigated. In this work, mats made of Polyvinylalcohol/Sodium alginate/Calendula officinalis with different percentages of calendula were prepared by electrospinning method to be used as wound dressings (Tahami et al.'s (2022).

The process of creating an alginate wound dressing involves a series of steps: (1) a mixture of sodium and calcium alginate is coated with ethyl alcohol to prevent the alginate from gelling in contact with water. (2) adding deionized or distilled water and obtaining an alginate solution, (3) impregnating, followed by drying a woven or nonwoven material with an alginate solution, (4) mechanical softening of the dressing to obtain a soft material and Flexible (Soorbaghi, et al.'s, 2019).

In addition to its ability to form a dressing, alginate can contain various natural substances that, when released together with calcium ions, can activate prothrombin and improve hemostasis.

Alginate wound dressings (Rezaie, M. et al.'s, 2021; Rezaie, M. et al.'s, 2022) absorb a large amount of liquid into the fiber structure in addition to the fluids from the fibers of the textile structure. Alginate wound dressing has antimicrobial and hemostatic properties, so it promotes wound healing. Given that infections lead to delays in the healing process, these features are useful. They are also widely used in the management of exuding wounds such as surgical wounds, leg ulcers, and pressure ulcers.

Currently, they are looking for solutions to replace synthetic polymers, which are widely used in medicine due to their inertness, high mechanical performance, and the fact that, unlike biopolymers, they can be easily moved and processed.

Film

In vitro permeation studies also showed that alginate film could be safely applied for contact delivery because a very small amount of mupirocin could pass through the epidermal layer, thereby accelerating the regeneration of the epidermal layer. In addition, compared to the commercial product, this film has better advantages for epithelialization, granulation tissue thickness, and angiogenesis. Due to the challenges of the new century, these natural materials have replaced synthetic polymers, ceramics, and metal alloys. Biomaterials are defined as materials intended to interface with biological systems to evaluate, treat, enhance, or replace any tissue, organ, or function of the body. To be used in the medical industry, biomaterials must be inert and not interact with the host organism. Therefore, natural materials have been widely used for this purpose, and the applications of replacing damaged tissues have been used and reported in specialized literature for a long time (Soorbaghi, et al.'s, 2019).

Foam Dressing

Currently, wound dressings have been widely replaced by foam dressings. Because the foam dressing is thicker than the wound dressing, it creates a better protective effect on the wound. They not only prevent the accumulation of exudate but also maintain hydration so that a suitable moist environment promotes epithelialization and healing by encouraging cell migration. Alginate foam dressings are made by intensively homogenizing a mixture of a water-soluble sodium alginate aqueous solution with a release agent, an emollient, and a surface-active agent, followed by the addition of a divalent or trivalent metal ion to form a water-insoluble material. We pour the resulting mixed hydrogel into a tray and put it in the freezer to obtain a frozen alginate hydrogel. The final step involves lyophilizing the mixture to remove moisture.

Alginate foam dressings can be enriched with active agents such as silver, which is used as an antimicrobial agent, which is an herbal wound healing agent. In this preparation, alginate produces the highest release of silver and asiaticoside, improves foam properties by increasing absorbency and compressive strength, has a comparable antimicrobial effect in the disc diffusion test, and shows non-cytotoxicity. Finally, the alginate dressing maintains a moist environment around the wound, regenerating the wound tissue, while ensuring the disappearance of granulation tissue and eliminating the possibility of tissue damage during wound debridement (Soorbaghi, et al.'s, 2019).

Hydrogel Dressing

Sodium alginate has been used to produce hydrogel dressings. Hydrogel dressings are of great interest due to their capacity to produce an ideal hydration environment for healing. To obtain the hydrogel dressing, sodium alginate, deionized water, and H2S are mixed and vortexed for one hour. CaCl2 is added and gently mixed to initiate ionic cross-linking of alginate polymer chains. Other cations such as divalent cations (Ca^{2+}, Sr^{2+}, Cd^{2+}, Zn^{2+}, Ba^{2+}, Cu^{2+}) or trivalent cations (Al^{3+}, Fe^{3+}) can lead to the formation of ionic crosslinked alginate hydrogels. The relationship between divalent cations and alginate chains using the "egg-box" model (external gelation, preferred for the synthesis of materials used in bone tissue engineering) and gelation mechanism (internal gelation, preferred for hydrogels) in situ, suitable for) is explained. Applications of alginate are injectable and are considered the most common and easiest way to encapsulate soluble and insoluble drugs. Alginate hydrogels can be chemically modified by altering molecular weight (alginates with higher molecular weights produce stiffer gels, while those with medium and low molecular weights allow for greater cell degradation and proliferation) (Gheorghita, P. et al.'s, 2020; Pahlevanzadeh, F. et al.'s; Dong et al.'s, 2013).

Physiological soft tissue is reassembled because it has low surface tension, oxygen permeability, and good mechanical and wetting properties. For these reasons, polysaccharides that exhibit hydrogel-forming properties are useful as wound dressing material. With cross-linked 3D polymer networks, high biocompatibility, and biodegradability, hydrogels are effectively used as controlled drug delivery and wound healing systems. To be effective, hydrogels must be sticky, elastic, and durable, and must be blocked and impermeable to bacteria. Alginate can form a hydrogel under very mild conditions at 25°C. However, no adhesive alginate hydrogel supports the growth of pluripotent stem cell-derived intestinal organoids, because the properties of alginate that support the growth of human intestinal organoids (HIOs) *in vitro* lead to epithelial differentiation of HIOs when The transplant takes place inside the body. To improve their properties, hydrogel films can be combined with various natural materials. For example, alginate/aloe vera hydrogels have been developed for wound healing applications, thus combining the hemostatic properties of calcium alginate gels with the healing properties of aloe vera. The results show that alginate/aloe vera hydrogel films can be evaluated as wound dressings for dry and oozing wounds (Rastogi, P. and Balasubramanian, K. 2019).

Cell Therapy

The biological properties of alginate, such as its interaction with cells by bioadhesive bonds, make it a promising biomaterial for cell and tissue culture. Alginates as 3D systems with ECM mimicking capacity have been used natively and are very useful for evaluating and understanding complex cell physiology, drug evaluation, and tissue engineering. Biomaterials used in cell therapy must have appropriate porosity, sufficient pore size, and interconnected pore structure to transport cells, metabolites, nutrients, and signaling molecules. Alginate biomaterials with macroporous structures create suitable conditions for cell attachment, proliferation, and differentiation for cell encapsulation with the ability to mimic the natural physiological environment. These 3D alginate constructs increase cell seeding efficiency.

According to the above, alginate is widely recognized as the most suitable polymer for cell encapsulation. Production of alginate scaffolds for cell encapsulation can be done by inkjet-based and extrusion-based systems. Cell encapsulation is a technique that involves immobilizing cells in a polymer gel and maintaining cellular metabolic activity. Encapsulated cells in alginate biomaterials grow and secrete new ECM and regenerate damaged tissue. This cutting-edge technology is important for advancing knowledge about cell physiology, drug delivery, tissue engineering, and regenerative medicine.

Also, alginate modifications with bioactive peptides have often been used as cell culture substrates *in vitro*. These biomaterials mimic the adhesive properties of ECM and stimulate cellular responses such as differentiation and proliferation. The presence of RGD peptides in alginate gels helps to control the cell phenotype of myoblasts, chondrocytes, osteoblasts as well as bone marrow stromal cells (BMSCs). For example, when RGD peptides are combined with alginate, the adhesion and proliferation of myoblasts cultured on alginate gels are enhanced.

In the last 10 years, research has been carried out on the development of cell transplantation therapy using ImmupelTM alginate encapsulation technology (LCT, Living Cell Technologies Limited, Australia) and commercial 3D alginate-based products for cell culture, such as AlgiMatrix (Dong et al.'s, 2013).

Conclusion and Perspectives

Recently, due to its controllable properties, gelation, simple modification to obtain alginate derivatives with different properties, ability to be implanted in various biomedical applications and high biocompatibility, alginate has attracted the attention of researchers in the fields of biomaterial and biotechnology. Due to these characteristics, alginate is used in many biomedical applications, including wound dressings, tissue engineering, drug delivery systems, etc. Alginate gels have limitations in terms of mechanical properties, which challenges researchers for a specific application with specific mechanical properties; However, researchers are trying to overcome this challenge with different strategies in the field of cross-linking and in combination with other suitable materials. According to the efforts of researchers to obtain the desired properties from species of bacteria

(Pseudomonas and Azotobacter) and brown seaweed sources and the subsequent modification of alginates, it seems that the use of alginates in biomedical applications will increase in the future, and the published articles will partially explain this increase.

References

Ahmed, Shakeel. (2019). Alginates: Applications in the Biomedical and Food Industries.

Andrzej, H., Kiryczyński, G., Marek, J. L., Chapter 7 - Biomaterials, Definition, Overview, Editor(s): Marek J. Łos, Andrzej Hudecki, Emilia Wiecheć, *Stem Cells and Biomaterials for Regenerative Medicine,* Academic Press, 2019, Pages 85-98.

Buddy, D. R., Hoffman, A. S., Schoen, F. J., Lemons, J. E., Wagner, W. R., Sakiyama-Elbert, S. E., Zhang, G. and Yaszemski, M. J. (2020). Introduction to Biomaterials Science: An Evolving, Multidisciplinary Endeavor, Editor(s): William R. Wagner, Shelly E. Sakiyama-Elbert, Guigen Zhang, Michael J. Yaszemski, *Biomaterials Science* (Fourth Edition), Academic Press, Pages 3-19.

Donati, I. and Paoletti, Sergio. (2009). *Material Properties of Alginates.* 10.1007/978-3-540-92679-5_1.

Dong, Qiu,Y., Chen, M. Y., Xin, Y., Qin, X. Y., Cheng, Z., Shi, L. E. and Tang, Z. X. (2013) Alginate-based and protein-based materials for probiotics encapsulation: a review. *International Journal of Food Science & Technology,* 48, no. 7: 1339-1351.

Farokhi, M., Jonidi Shariatzadeh, F., Solouk, A. and Mirzadeh, H. Alginate based scaffolds for cartilage tissue engineering: a review. *International Journal of Polymeric Materials and Polymeric Biomaterials,* 69, no. 4: 230-247.

Gheorghita, P., Roxana, A. L., Mihai, D., and Mihai, C. (2020) Alginate: From food industry to biomedical applications and management of metabolic disorders. *Polymers,* 12, no. 10: 2417.

Hasnain, M. D., Jameel, E., Mohanta, B., Dhara, A., Alkahtani, S. and Nayak, Amit. (2020). Alginates: sources, structure, and properties.

Hasnain, M. D., Nayak, A., Yadav, M., Ahmadi, Y., Milivojevic, M., & Pajić-Lijaković, I., Branko, B., Manzano, V., Pacho, N., Tasqué, J., Beatriz, N., Accorso, D., Singh, V., Singh, A., El-Sherbiny, I., Abd, M., Aziz, A., Abdelsalam, E., Garcia, N., and Dmour, I. (2019). Alginates Versatile Polymers in Biomedical Applications and Therapeutics Chemically Modified Alginates for Advanced Biomedical Applications.

Migonney, V. (2014). Definitions. In Biomaterials, V. Migonney (Ed.).

Rezaei, M., Hassanzadeh Nemati, N., Mehrabani, D. and Komeili, Ali. (2021) Skin Regeneration by Hybrid Carboxyl Methyl Cellulose/Calcium Alginate Fibers Electrospun Scaffold, *Journal of Natural Fibers,.*

Rezaei, M., Hassanzadeh Nemati, N., Mehrabani, D. and Komeili, Ali. (2022). Characterization of sodium carboxymethyl cellulose/calcium alginate scaffold loaded with curcumin in skin tissue engineering. *Journal of Applied Polymer Science,* 139(22).

Moore, A. (2015). Alginic acid: Chemical structure, uses and health benefits.

Moradali, M., Ghods, S. and Rehm, B. (2018). Alginate Biosynthesis and Biotechnological Production. 978-981.

Pahlevanzadeh, F., H Mokhtari, A., Bakhsheshi-Rad, H., Emadi, R., Kharaziha, M., Valiani, Ali., Poursamar, Ali., Fauzi Ismail, A., RamaKrishna, S. and Berto, F. Recent trends in three-dimensional bioinks based on alginate for biomedical.

Pereira, L. and Cotas, J. (2020). Alginates - A General Overview. 10.5772/*intechopen*, 88381.

Rastogi, P. and Balasubramanian Kandasubramanian. (2019). Review of alginate-based hydrogel bioprinting for application in tissue engineering. *Biofabrication*, 11, no. 4: 042001.

Reddy, S. (2021). Alginates - A Seaweed Product: Its Properties and Applications In Properties and Applications of Alginates, edited by Irem D., Esra I. and Tugba Keskin-Gundogdu. London: IntechOpen.

Rinaudo, M. (2013). Biomaterials based on a natural polysaccharide: Alginate. TIP. *Revista especializada en ciencias químico-biológicas*, 17. 92-96.

Saji, S., Hebden, A., Goswami, P. and Du, C. (2022). A Brief Review on the Development of Alginate Extraction Process and Its Sustainability. *Sustainability*, 14. 5181.

Soorbaghi, Pashaei, F., Isanejad, M., Salatin, S., Ghorbani, M., Jafari, S. and Derakhshankhah, H. (2019) Bioaerogels: Synthesis approaches, cellular uptake, and the biomedical applications. *Biomedicine & Pharmacotherapy*, 111: 964-975.

Sun, J. and Huaping, T. (2013). Alginate-Based Biomaterials for Regenerative Medicine Applications *Materials* 6, no. 4: 1285-1309.

Szekalska, M., Puciłowska, A., Szymańska, E., Ciosek, P. and Winnicka, K. (2016). Alginate: Current Use and Future Perspectives in Pharmaceutical and Biomedical Applications. *International Journal of Polymer Science*, 1-17.

Tahami, S. R., Hasanzadeh Nemati, N., Keshvari, H. and Khorasani, M. T. (2020). Effect of Electrical Potential on the Morphology of Polyvinyl Alcohol/Sodium Alginate Electrospun Nanofibers, Containing Herbal Extracts of Calendula Officinalis for Using in Biomedical Applications. J*ournal of Modern Processes in Manufacturing and Production*, 9(2), 43-56.

Tahami, S. R., Hassanzadeh Nemati, N., Keshvari, H. and Khorasani, M. T. (2022). *In vitro* and *in vivo* evaluation of nanofibre mats containing Calendula officinalis extract as a wound dressing. *Journal of Wound Care*, 31:7, 598-611.

Zhang, H., Cheng, J. and Ao, Q. (2021). Preparation of Alginate-Based Biomaterials and Their Applications in Biomedicine. *Mar Drugs*, 19(5), 264.

Index

A

alginate gels, 8, 10, 12, 22, 32, 35, 78, 82, 83, 88, 90, 92, 95, 157, 159, 160, 161, 162, 168, 170, 171, 175, 176
alginate oligomers, 19
alginate sulfate, 32, 42, 66
alginic acid, 4, 6, 7, 8, 9, 15, 25, 27, 30, 31, 77, 82, 87, 100, 104, 124, 152, 153, 166, 167, 168
animal feed, vii, 1, 85, 90
applications, v, vi, vii, viii, ix, x, 1, 2, 11, 15, 16, 17, 19, 20, 22, 23, 25, 26, 29, 31, 33, 34, 41, 42, 43, 64, 65, 66, 67, 68, 69, 70, 72, 75, 76, 77, 79, 81, 83, 85, 86, 87, 91, 92, 95, 96, 97, 98, 99, 100, 103, 107, 119, 122, 123, 124, 125, 135, 147, 148, 149, 151, 154, 159, 162, 165, 167, 168, 170, 172, 174, 175, 176, 177, 178
azotobacter, x, 2, 4, 28, 29, 77, 107, 133, 165, 166, 167, 168, 177

B

beads, v, vii, ix, 14, 17, 23, 24, 27, 64, 88, 89, 93, 95, 97, 98, 99, 103, 104, 105, 106, 108, 109, 110, 111, 112, 113, 114, 115, 116, 117, 118, 119, 120, 121, 122, 123, 124, 125, 126, 129, 131, 133, 134, 135, 137, 138, 139, 140, 141, 142, 146, 148, 149, 153, 158, 159, 163
bioadhesion, 81, 87, 98
biocompatibility, vii, viii, ix, x, 1, 10, 16, 18, 28, 33, 35, 42, 75, 76, 81, 84, 92, 96, 100, 108, 125, 131, 151, 154, 158, 163, 167, 169, 175, 176
biocompatible, ix, x, 17, 67, 75, 93, 103, 104, 112, 135, 151, 162, 166, 167
biodegradability, viii, 16, 22, 75, 76, 108, 151, 169, 175

C

carbodiimide, 12, 14, 46, 47
carboxyl group, 11, 13, 33, 81, 106, 140
cardiovascular disease(s), 17
cosmetic(s), vii, viii, x, 1, 2, 26, 75, 76, 85, 89, 154, 163, 165, 170
covalent cross-linking, 12, 157, 170
cross-linking, ix, 7, 18, 42, 55, 78, 83, 94, 96, 101, 106, 139, 151, 158, 159, 160, 169, 174, 176

D

derivatives of alginate, 2
drug delivery systems, x, 17, 35, 79, 94, 97, 100, 101, 104, 125, 154, 159, 165, 166, 176

E

egg box model, 9, 75, 83
encapsulation, v, vii, viii, 10, 21, 25, 32, 35, 41, 42, 43, 44, 52, 57, 62, 67, 68, 77, 97, 99, 101, 104, 106, 107, 109, 110, 112, 113, 114, 115, 116, 117, 118, 119, 120, 121, 122, 123, 124, 125, 128, 129, 133, 135, 136, 137, 141, 142, 143, 144, 145, 146, 147, 148, 149, 150, 158, 160, 163, 170, 176, 177
extrusion, v, vii, ix, 32, 64, 91, 109, 121, 124, 125, 131, 132, 133, 134, 135, 136, 137, 138, 142, 143, 146, 147, 176

F

food industry, vii, ix, 1, 10, 21, 26, 91, 103, 104, 106, 107, 109, 112, 119, 123, 124, 131, 133, 135, 136, 147, 177 functional food(s), 20, 112, 119, 124, 126, 127, 132, 135, 147, 148, 149, 150

G

gastrointestinal, ix, 19, 40, 87, 108, 112, 113, 114, 115, 122, 123, 131, 132, 133, 138, 141, 142, 145, 146, 149, 160
gel particle(s), v, vii, viii, 41, 42, 44, 45, 48, 50, 53, 54, 58, 60, 62, 64, 121, 159, 171
gelling properties, 83, 108, 156
graft copolymerization, 14
grafting, 12, 14, 73, 74, 159, 171

H

heparin (HEP), viii, 14, 29, 32, 41, 42, 43, 45, 51, 52, 53, 63, 64, 65, 66, 67, 157, 160, 161
hydrogels, x, 8, 12, 16, 17, 18, 29, 30, 31, 33, 34, 35, 65, 66, 71, 84, 88, 90, 91, 92, 93, 95, 96, 97, 101, 112, 119, 120, 122, 123, 124, 126, 127, 128, 129, 134, 137, 149, 151, 157, 158, 161, 162, 165, 166, 169, 172, 174, 175

I

immobilization, ix, 10, 14, 23, 24, 32, 33, 66, 87, 97, 99, 100, 131, 132, 133, 147, 170
immunogenicity, 84, 167, 169
isolation, 30, 46, 48, 49, 50, 54, 55, 56, 58, 65, 66, 70, 73

L

Lactoferrin (Lf), viii, 41, 42, 43, 44, 45, 47, 48, 49, 51, 52, 53, 54, 55, 56, 57, 58, 59, 60, 61, 62, 63, 64, 66, 67, 68

M

metabolic syndrome, 19
molecular weight, 6, 8, 14, 15, 19, 27, 32, 34, 35, 52, 54, 76, 77, 82, 83, 84, 85, 87, 95, 105, 117, 123, 159, 163, 169, 175

N

nanoprecipitation, viii, 41, 43, 44, 48, 49, 54, 55, 56, 58, 59, 60, 62

O

oxidation, 8, 13, 29, 30, 33, 113, 145, 163

P

particle size distribution, 30, 42, 50, 58, 62, 63
particle size distribution and suspension stability, 42
pH values, 7, 20, 82, 105
pharmaceutical application, 87, 94, 156, 171
pharmaceutical industry(ies), viii, 26, 75, 86, 154, 156, 158, 167, 170
pharmaceuticals, viii, 75, 76, 85, 94, 156, 157, 159, 169
phosphorylation, 15, 28, 128
polymers, v, viii, ix, 2, 4, 8, 13, 14, 16, 17, 26, 27, 29, 30, 33, 34, 35, 36, 38, 39, 40, 42, 47, 53, 54, 56, 64, 65, 66, 71, 73, 75, 76, 81, 84, 90, 92, 95, 96, 97, 98, 99, 112, 113, 121, 122, 124, 135, 148, 151, 158, 162, 169, 170, 171, 172, 173, 177
polyphenols, v, ix, 103, 104, 105, 112, 113, 114, 115, 116, 117, 119, 120, 121, 122, 123, 124, 125, 126, 128, 129
polysaccharide, ix, 2, 14, 23, 32, 35, 42, 52, 65, 67, 74, 76, 86, 88, 89, 98, 105, 108, 124, 131, 133, 152, 154, 167, 168, 169, 178

probiotic, v, vii, ix, 31, 107, 124, 131, 132, 133, 134, 136, 137, 138, 140, 141, 142, 143, 144, 145, 147, 148, 149, 150

probiotic encapsulation, 134, 137, 138, 141, 142, 143, 147, 149

probiotic immobilization, v, vii, ix, 131, 137, 143

properties, v, vi, vii, viii, ix, x, 1, 2, 6, 8, 10, 14, 15, 16, 18, 19, 22, 26, 27, 28, 30, 31, 32, 33, 34, 35, 36, 37, 38, 43, 45, 63, 64, 65, 67, 74, 75, 76, 77, 79, 80, 81, 83, 86, 87, 94, 95, 96, 97, 98, 99, 101, 103, 104, 105, 106, 107, 108, 109, 112, 116, 119, 121, 122, 123, 124, 127, 132, 133, 135, 137, 138, 139, 140, 141, 142, 144, 145, 146, 147, 148, 149, 150, 151, 154, 156, 158, 165, 167, 168, 169, 171, 172, 173, 174, 175, 176, 177, 178

protein delivery, 26, 88, 157, 158, 164

S

seaweed, x, 2, 3, 4, 6, 28, 31, 34, 76, 82, 99, 153, 154, 155, 163, 165, 166, 167, 169, 177, 178

selective ion binding, 9

solubility, 7, 30, 42, 45, 78, 79, 82, 85, 106, 112, 159, 169, 171

sonication, viii, 42, 43, 44, 48, 49, 54, 55, 56, 57, 59, 60, 61, 62

spray dry, v, vii, ix, 125, 131, 132, 135, 136, 143, 144, 145, 146, 147, 148, 150

sterilization, 85, 100

sulfated alginate (S-ALG), v, vii, viii, 14, 41, 42, 43, 44, 45, 46, 47, 48, 49, 51, 52, 53, 54, 55, 56, 57, 58, 62, 63

sulfation, 14, 29, 45, 54, 65

surface plasmon resonance, viii, 41, 42, 45, 65

suspension stability, 45, 50, 58, 59, 62

T

textile(s), vii, 1, 23, 26, 27, 35, 85, 90, 98, 154, 155, 170, 173

tissue engineering, vii, x, 16, 29, 30, 35, 66, 70, 72, 79, 82, 83, 96, 100, 161, 162, 165, 166, 170, 171, 172, 174, 175, 176, 177, 178

toxicity, vii, x, 42, 85, 151, 154, 158, 167, 169, 170

tributyl ammonium salt of ALG (T-ALG), 45, 46

V

viscosity, 6, 7, 15, 20, 44, 45, 53, 77, 82, 84, 85, 86, 89, 91, 104, 105, 110, 111, 123, 152, 155, 167, 169

W

welding, viii, 26, 75, 85, 91, 155

wound dressings, x, 8, 18, 26, 89, 90, 160, 163, 165, 172, 173, 174, 175, 176

α

α-L-guluronic acid, vii, 1, 2, 23, 75, 105, 168

β

β-cyclodextrin, 12, 66

β-D-mannuronic acid, vii, 1, 2, 23, 75, 105, 168